T0138170

The ADI Model Problem

Eugene Wachspress

The ADI Model Problem

Eugene Wachspress
East Windsor, New Jersey
USA

Additional material to this book can be downloaded from http://extras.springer.com

ISBN 978-1-4939-0019-0 ISBN 978-1-4614-5122-8 (eBook)
DOI 10.1007/978-1-4614-5122-8
Springer New York Heidelberg Dordrecht London

© Springer Science+Business Media New York 2013
Softcover re-print of the Hardcover 1st edition 2013
This work is subject to copyright. All rights are reserved by the Publisher, whether the whole or part of the material is concerned, specifically the rights of translation, reprinting, reuse of illustrations, recitation, broadcasting, reproduction on microfilms or in any other physical way, and transmission or information storage and retrieval, electronic adaptation, computer software, or by similar or dissimilar methodology now known or hereafter developed. Exempted from this legal reservation are brief excerpts in connection with reviews or scholarly analysis or material supplied specifically for the purpose of being entered and executed on a computer system, for exclusive use by the purchaser of the work. Duplication of this publication or parts thereof is permitted only under the provisions of the Copyright Law of the Publisher's location, in its current version, and permission for use must always be obtained from Springer. Permissions for use may be obtained through RightsLink at the Copyright Clearance Center. Violations are liable to prosecution under the respective Copyright Law.
The use of general descriptive names, registered names, trademarks, service marks, etc. in this publication does not imply, even in the absence of a specific statement, that such names are exempt from the relevant protective laws and regulations and therefore free for general use.
While the advice and information in this book are believed to be true and accurate at the date of publication, neither the authors nor the editors nor the publisher can accept any legal responsibility for any errors or omissions that may be made. The publisher makes no warranty, express or implied, with respect to the material contained herein.

Printed on acid-free paper

Springer is part of Springer Science+Business Media (www.springer.com)

In loving memory of my wife of 53 years
Natalie

Preface

This work is an updated edition of my self-published monograph on The ADI Model Problem [Wachspress, 1995]. Minor typographic corrections have been made in Chaps. 1–4. A few innovations have been added such as non-iterative alignment of complex spectra in Sect. 4.6. In 1995 the theory was pleasing but application was limited. With regard to discretized elliptic equations, the restriction to rectangular grids for which reasonable separable approximations could be generated was particularly limiting. Lyapunov and Sylvester matrix equations opened new applications. Although the commutation property was no longer a problem there were limiting factors here too. Accurate estimates for crucial eigenvalues were required in order to find effective iteration parameters. This led to a study described in Sect. 3.8 in Chap. 3 of similarity reduction to banded upper Hessenberg form which ended with "In time, an efficient and robust scheme will emerge for application to the Lyapunov problem." Such a scheme is presented in Chap. 5 of this edition. Crucial eigenvalues include those with small real part, those which subtend large angles at the origin and that of largest magnitude. When the reduced matrix is expressed in sparse form the MATLAB program EIGS provides precisely the tool needed for this determination.

This left only the limiting fact that ADI, although competitive with methods like Bartels–Stewart [Bartels, 1972], did not offer enough advantage to supplant standard methods. Lack of familiarity of practitioners with the new theory was also detrimental. The saving innovation was that of Penzl [Penzl, 1999]. He demonstrated that the ADI approach allowed one to reduce arithmetic significantly for problems with low-rank right-hand sides. This was not possible with the other schemes. Further analysis by Li and White [Li and White, 2002] reduced the arithmetic even further. Now the ADI approach was worthy of consideration by practitioners. It is my hope that it will become the method of choice not only for low-rank problems but also for most Lyapunov and Sylvester problems.

My programs are all written for serial computation. The ADI approach parallelizes readily. I have described in Sect. 5.7 a scheme for approximating a full symmetric matrix by a sum of low-rank matrices. Each of the low-rank matrices may be used as a right-hand side in parallel solution of low-rank Lyapunov equations. I only discuss symmetric right-hand sides in Sect. 5.7. Sylvester equations have nonsymmetric right-hand sides. Biorthogonal Lanczos schemes extend to nonsymmetric matrices. Lanczos difficulties associated with near-zero inner products may be preempted in this approach by restart to find the next low-rank component.

In 1995 I was still writing FORTRAN programs. I now work exclusively with MATLAB. Chapter 6 now includes MATLAB implementation of theory described in Chaps. 1–5. These programs have treated successfully all my test problems. Theory was supported in that when crucial eigenvalues were computed accurately the observed error reduction agreed with prediction.

This work describes my effort over the past 50 odd years related to theory and application of the ADI iterative method. Although ADI methods enjoyed widespread use initially, model problem conditions were not present and little use was made of the theory related to elliptic functions. The more recent relevance to Lyapunov and Sylvester equations stimulated the first significant application. I am hopeful that this will lead to further analysis and that other areas of application will be uncovered. That the simply stated minimax problem defined in Eqs. 8–10 in Chap. 1 could lead to the theory exposed in this work has never ceased to amaze me.

East Windsor, NJ Eugene Wachspress

Contents

Chapter 1
The Peaceman–Rachford Model Problem

Abstract Early analysis of ADI iteration introduced by Peaceman and Rachford led to application of Chebyshev minimax theory to determine optimal parameters. Elliptic functions play a crucial role. It was discovered belatedly that this problem had been solved in 1877 by Zolotarev.

1.1 Introduction

Alternating-direction-implicit (ADI) iteration was introduced in [Peaceman and Rachford, 1955] as a method for solving elliptic and parabolic difference equations. There has been a wealth of analysis and application since that time. In this work, theory for and practical use of the so-called model problem will be examined in a more general context than initially considered. The original Peaceman–Rachford formulation and early generalizations will be described first. An excellent survey of early ADI theory may be found in [Birkhoff, Varga and Young, 1962]. Much of this early theory is developed in [Wachspress, 1966].

Let A be a real symmetric positive-definite (SPD) matrix of order n and let $\mathbf{s}$ be a known n-vector. A recurring problem in linear algebra is to find the vector $\mathbf{u}$ which solves the linear equation

$$A\mathbf{u} = \mathbf{s}. \tag{1}$$

Let I be the identity matrix of order n. Then ADI iteration can be applied when A can be expressed as the sum of matrices H and V, for which the linear systems $(H + pI)\mathbf{v} = \mathbf{r}$ and $(V + pI)\mathbf{w} = \mathbf{t}$ admit an efficient solution. Here, p is a suitably chosen iteration parameter and $\mathbf{r}$ and $\mathbf{t}$ are known (see Eq. 2).

If H and V are both SPD, then there are positive parameters p_j for which the two-sweep iteration defined by

$$(H + p_j I)\mathbf{u}_{j-\frac{1}{2}} = (p_j I - V)\mathbf{u}_{j-1} + \mathbf{s}, \tag{2.1}$$

$$(V + p_j I)\mathbf{u}_j = (p_j I - H)\mathbf{u}_{j-\frac{1}{2}} + \mathbf{s} \tag{2.2}$$

E. Wachspress, *The ADI Model Problem*, DOI 10.1007/978-1-4614-5122-8_1,
© Springer Science+Business Media New York 2013

for $j = 1, 2, \ldots$ converges. Here, $\mathbf{u}_0$ is prescribed and is often taken as the **0**-*vector*. Equation 2 are the PR ADI (Peaceman–Rachford Alternating Direction Implicit) iteration equations. In Peaceman and Rachford's application, H was the difference approximation at nodes of a rectangular grid of the negative of the second derivative with respect to x and V was the difference approximation at these same nodes of the negative of the second derivative with respect to y.

Definitive analysis has been applied to the special case where matrices H and V commute. However, the method is often used when matrices H and V do not commute.[1] Then convergence theory is limited and even with optimum iteration parameters PR ADI may not be competitive with other methods. When H and V commute convergence is more rapid (when readily computed optimum parameters are used) than other methods in current use, this led to the definition of the model ADI problem as one where A is SPD and the SPD matrices H and V commute [Wachspress, 1957].

It was also observed that the iteration equations could be generalized [Wachspress, 1963] by introduction of an SPD normalization matrix, F, and furthermore that when the spectra of H and V differed an improvement in efficiency could be realized by changing the iteration parameter each sweep. The Peaceman–Rachford equations were thus generalized to

$$(H + p_j F)\mathbf{u}_{j-\frac{1}{2}} = (p_j F - V)\mathbf{u}_{j-1} + \mathbf{s}, \tag{3.1}$$

$$(V + q_j F)\mathbf{u}_j = (q_j F - H)\mathbf{u}_{j-\frac{1}{2}} + \mathbf{s}. \tag{3.2}$$

The commutation property crucial for definitive analysis of these equations is $HF^{-1}V - VF^{-1}H = 0$. One example of the importance of F is in solution of Poisson's equation with variable mesh increments over a rectangular grid. Let $\lambda(F^{-1}H)$ and $\gamma(F^{-1}V)$ be eigenvalues of $F^{-1}H$ and $F^{-1}V$. Then the model-problem condition for Eq. 2 that H and V both be SPD is generalized to $\lambda(F^{-1}H) + \gamma(F^{-1}V) > 0$. This will be clarified subsequently as will the importance of using a different parameter on each sweep.

The elegance of the underlying theory and the rapid convergence rate for the model problem led to search for such problems. Three major applications emerged. First, there are systems which naturally lead to model ADI problems. These include solution of Poisson's equation over a rectangle. Second, there are systems which are closely related to model problems. To solve a related system, one may use a model problem as a preconditioner and couple an "inner" model-problem ADI iteration with an "outer" iteration on the deviation of the true problem from the model problem [Cesari, 1937; Wachspress, 1963]. The method now known as preconditioned conjugate gradients (PCCG) was introduced for solution of elliptic systems in this context.

[1] Not long after Peaceman and Rachford introduced the *ADI* method, John Sheldon (then at Computer Usage Co.) described to me why rapid convergence rates may be ensured only when matrices H and V commute.

Third, there is a class of problems which arise in an entirely different context which are model ADI problems. This latter class is the Sylvester matrix equation $AX + XB = C$ and the special case of the Lyapunov matrix equation $AX + XA^\top = C$. In these equations, A, B, and C are given matrices and X is to be found. Applicability of ADI iteration to these problems was disclosed 30 years after the seminal Peaceman and Rachford work and has stimulated generalization of ADI theory from real to complex spectra [Ellner and Wachspress, 1986, 1991; Saltzman, 1987; Wachspress, 1988, 1990; Starke, 1988; Lu and Wachspress, 1991; Istace and Thiran, 1993].

The initial ADI equations were applied to five-point differencing of Laplacian-type operators. Finite-element discretization over rectangular grids with bilinear basis functions leads to nine-point rather than five-point equations. It was discovered [Wachspress, 1984] 30 years after the introduction of ADI for five-point equations that one could devise a generalization to the nine-point equations of the finite-element method. Prior to this discovery, five-point ADI had been applied as a preconditioner to solve nine-point equations. The discovery of nine-point ADI equations eliminated the outer iteration of this earlier method for separable Laplacian-type finite-element equations over rectangles and provided a better model problem for solution of nonseparable problems of this type.

It is apparent from this introductory discussion that ADI iteration has been a fertile area for numerical analysis since its inception. As is often true in diverse fields of mathematics, just when one seems to have concluded a line of research a new outlook uncovers an unexpected continuation.

1.2 The Peaceman–Rachford Minimax Problem

Convergence theory was first developed for the Peaceman–Rachford iteration of Eq. 2 and subsequently for the more general Eq. 3. Let the error in iterate j be denoted by $\mathbf{e}_j = \mathbf{u}_j - \mathbf{u}$. Then Eq. 2 yield $\mathbf{e}_j = R_j\ \mathbf{e}_{j-1}$, where

$$R_j \stackrel{\text{def}}{=} (V + p_j I)^{-1}(H - p_j I)(H + p_j I)^{-1}(V - p_j I). \tag{4}$$

The commuting SPD matrices H and V have a simultaneous eigenvector basis which is also a basis for R_j and is independent of the iteration parameter p_j. Let λ and γ be eigenvalues of H and V, respectively. In application to five-point difference equations, there is an eigenvector which has any of the λ and γ pairs as respective eigenvalues. The error after J iterations satisfies $\mathbf{e}_J = G_J\ \mathbf{e}_0$, where

$$G_J \stackrel{\text{def}}{=} \prod_{j=1}^{J} R_j. \tag{5}$$

Unless stated otherwise, the symbol $\|X\|$ denotes the 2-norm of X and $\rho(X)$ denotes the spectral radius of X. Commutation of H and V assures symmetry of G_J, and it follows that $\|G_J\| = \rho(G_J)$. Hence, $\|e_J\| \leq \rho(G_J)\,\|e_0\|$, where

$$\rho(G_J) = \max_{\lambda\,,\,\gamma} \left| \prod_{j=1}^{J} \frac{(p_j - \lambda)(p_j - \gamma)}{(p_j + \lambda)(p_j + \gamma)} \right|. \tag{6}$$

The minimax problem for the Peaceman–Rachford ADI equations is for a given J to choose a set of iteration parameters $\{p_j\}$ which minimizes $\rho(G_J)$. Although Peaceman and Rachford (1955), Douglas (1961), and Wachspress (1957) found reasonably efficient parameters for early application, the first truly optimum parameters were exposed in 1962 by Wachspress and independently by Gastinel for $J = 2^n$ and finally for all J by W.B. Jordan [Wachspress, 1963]. Successive stages of this development will now be traced.

Let $a \leq \lambda \leq b$ and $c \leq \gamma \leq d$. Only when the iteration is generalized to Eq. 3 can different ranges for λ and γ be treated properly. In the early analysis, the smaller of a and c was chosen as a lower bound and the larger of b and d as an upper bound for both spectra. Let these bounds be denoted by a and b in this analysis of parameter optimization for Eq. 2. When the entire spectrum is known, it may be possible to determine optimum parameters for the discrete spectrum. We defer discussion of this for the present. Even when this is possible, there is often little improvement over parameters chosen to solve the minimax problem under the assumption that λ and γ vary continuously over $[a, b]$. Moreover, in most applications, only eigenvalue bounds are known. The Peaceman–Rachford minimax problem for the spectral interval [a,b], assuming continuous variation of the eigenvalues, is

$$\min_{\{p_j\}} \max_{a\leq\lambda,\gamma\leq b} \rho(G_J). \tag{7}$$

Since λ and γ values can occur in any combination, they may be treated as independent variables and the minimax problem reduces to

$$\min_{\{p_j\}} \max_{a\leq x\leq b} |Q_J(x, \mathbf{p})|, \tag{8}$$

where $\mathbf{p}$ designates a J-tuple of real numbers and

$$Q_J(x, \mathbf{p}) = \prod_{j=1}^{J} \left(\frac{p_j - x}{p_j + x} \right). \tag{9}$$

Let

$$H(\mathbf{p}) = \max_{x\in[a,b]} |Q_J(x, \mathbf{p})|. \tag{10}$$

It should be noted that $\rho(G_J) = H^2(\mathbf{p})$. Existence of a solution is established by the usual compactness argument. If a value for p_j less than a is replaced by a or a value greater than b is replaced by b, all factors in Q decrease in absolute value over the entire interval $[a,b]$ (except where they remain zero.) Therefore, $\max|Q|$ is a bounded (obviously less than unity) continuous function of its arguments over the closed $(J+1)$-dimensional interval $[a,b]^{J+1}$ and hence uniformly continuous for the p_j and x in that interval. Hence, $\max|Q_J(x,\mathbf{p})|$ for $x \in [a,b]$ attains its minimum for $\mathbf{p}$ in the J-dimensional interval $[a,b]^J$.

An "alternance" property was disclosed by de La Vallée Poussin [Achieser, 1967] to play a crucial role in resolving minimax problems of this type. This property is the content of the following theorem:

Theorem 1 (The de La Vallée Poussin Alternance Property). *If $Q_J(x,\mathbf{p}_s)$ assumes the values $r_1, -r_2, r_3, -r_4, \ldots (-1)^J r_{J+1}$ with all the $r_j > 0$ at monotonically increasing values of $x \in [a,b]$, and if Q is continuous in $[a,b]$, then*

$$H \equiv \min_{\mathbf{p}} H(\mathbf{p}) = \min_{\mathbf{p}} \max_{x\in[a,b]} |Q_J(x,\mathbf{p})|$$

is bounded below by the smallest r_j.

Proof. Suppose there is a $\mathbf{p}_k$ such that

$$H_k < \min\ r_j. \tag{11}$$

It will be shown that this leads to a contradiction. The difference function

$$R_J(x) = Q_J(x,\mathbf{p}) - Q_J(x,\mathbf{p}_k)$$
$$= \frac{\prod_{j=1}^{J}(p_j - x)\left(p_j^{(k)} + x\right) - \prod_{j=1}^{J}(p_j + x)\left(p_j^{(k)} - x\right)}{\prod_{k=1}^{J}(p_j + x)\left(p_j^{(k)} + x\right)} \tag{12}$$

is a ratio of a polynomial of maximal degree $2J-1$ in x and a polynomial of degree $2J$ in x. The denominator is positive over $[a,b]$, and hence $R_J(x)$ is continuous over this interval. It follows from Eq. 11 that $R_J(x)$ has the sign of $Q_J(x,\mathbf{p})$ at the $J+1$ alternation points and must vanish at J intermediate points in $[a,b]$. The numerator polynomial is an odd function of x and must therefore also vanish at the corresponding points in $[-b,-a]$. This polynomial of maximal degree $2J-1$ has been shown to have at least $2J$ roots, which is not possible.

A lemma based on the Euclidean division algorithm for polynomials facilitates further analysis of this minimax problem.

Lemma 2 (Divisor Lemma). *Let $R_J(x)$ and $R_J(-x)$ be relatively prime polynomials of degree J (i.e., they have no common polynomial divisors other than constants). For any positive values for the x_i and any $t \le J$, define*

$$\phi(x) = x \prod_{i=1}^{t-1} (x_i - x)(x_i + x). \tag{13}$$

Then a polynomial $P_s(x)$ of degree $s < J$ may be found such that

$$\phi(x) \equiv P_s(x) R_J(-x) - P_s(-x) R_J(x). \tag{14}$$

Proof. The polynomial division algorithm establishes the existence of polynomials $P_s(x)$ and $Q_s(x)$ for which

$$\phi(x) \equiv P_s(x) R_J(-x) - Q_s(x) R_J(x). \tag{15}$$

Since $\phi(x) + \phi(-x) = 0$, we obtain from Eq. 15

$$R_J(x)[P_s(-x) - Q_s(x)] + R_J(-x)[P_s(x) - Q_s(-x)] = 0. \tag{16}$$

By hypothesis, $R_J(x)$ vanishes at J nonzero points, say x_j where $R_J(-x)$ is not zero. It follows from Eq. 16 that $P_s(x_j) - Q_s(-x_j) = 0$. Also by Eq. 16, $P_s(0) - Q_s(0) = 0$. These $J + 1$ roots of $P_s(x) - Q_s(-x)$, which is of maximal degree $J-1$, establish that this is the zero polynomial or that $Q_s(x) = P_s(-x)$. The lemma is thus proved.

This sets the stage for proof of the Chebyshev alternance property of the solution to the ADI minimax problem.

Theorem 3 (Chebyshev Alternance Theorem). *The function $Q_J(x, \mathbf{p}^o)$ having least deviation from zero in $[a, b]$ assumes its maximum magnitude, H, $J + 1$ times in $[a, b]$ with alternating sign.*

Proof. If Q_J assumes its maximum deviation $J + 1$ times in $[a, b]$ with alternating sign, then it follows from the de La Vallée Poussin theorem that this maximum must be the least possible value. It remains to be shown that such a function exists. Suppose for some $\mathbf{p}_s$, the number of alternation points at which $|Q_J(x, \mathbf{p}_s)| = H_s$ is equal to t where $t < (J+1)$. Let x_i for $i = 1, 2, \ldots, (t-1)$ separate the alternation points and let $x_0 = a$ and $x_t = b$. Then one of the following two inequalities holds in each interval $[x_i, x_{i+1}]$ for $i = 0, 1, 2, \ldots, (t-1)$ and some sufficiently small positive value for α:

$$-H_s \leq Q_J(x, \mathbf{p}_s) < H_s - \alpha, \tag{17.1}$$

$$-H_s + \alpha < Q_J(x, \mathbf{p}_s) \leq H_s. \tag{17.2}$$

Let ϕ in Eq. 13 be constructed with this set of x_i, excluding the a and b endpoints (i.e., $i = 0$ and $i = t$). Note that the degree of ϕ is equal to $2t - 1$ and that this is less than $2J$. Further, let the polynomial $R_J(x)$ be defined by $\mathbf{p}_s$ as

$$R_J(x) = \prod_{j=1}^{J} \left(p_j^{(s)} + x \right). \tag{18}$$

The divisor lemma establishes the existence of a polynomial $P_s(x)$ of maximal degree $2t - 1 - J$, which is less than J, satisfying Eq. 14. This polynomial can be used to construct a $\mathbf{p}_k$ for which $H_k < H_s$. For a sufficiently small β we obtain $\mathbf{p}_k$ from

$$Q_J(x, \mathbf{p}_k) = \frac{\prod_{j=1}^{J} \left(p_j^{(s)} - x \right) - \beta P_s(-x)}{\prod_{j=1}^{J} \left(p_j^{(s)} + x \right) - \beta P_s(x)} = \frac{\prod_{j=1}^{J} \left(p_j^{(k)} - x \right)}{\prod_{j=1}^{J} \left(p_j^{(k)} + x \right)}. \tag{19}$$

It is easily demonstrated by elementary algebraic manipulation that

$$Q_J(x, \mathbf{p}_k) = Q_J(x, \mathbf{p}_s) + \frac{\beta \phi(x)}{\prod_{j=1}^{J} \left(p_j^{(s)} + x \right) \left[\prod_{j=1}^{J} \left(p_j^{(s)} + x \right) - \beta P_s(x) \right]}. \tag{20}$$

For sufficiently small β, the denominator of the last term is positive over $[a, b]$ so that this term has the sign of $\beta \phi(x)$. By Eq. 13, $\beta \phi(x)$ alternates in sign in the successive intervals $[x_i, x_{(i+1)}]$ so that the sign of β may be chosen so that the inequalities in Eq. 17 become for $Q_J(x, \mathbf{p}_k)$:

$$-H_s - \theta + \alpha \le Q_J(x, \mathbf{p}_k) \le H_s - \theta, \tag{21.1}$$

$$-H_s + \theta \le Q_J(x, \mathbf{p}_k) \le H_s - \alpha + \theta, \tag{21.2}$$

where θ is some positive value which can be made closer to zero than α by choosing a small enough β. We may in fact choose β so that $\theta = \alpha/2$. This will yield $H_k \le H_s - \alpha/2$. The construction may be repeated until there are exactly $J + 1$ successive points in $[a, b]$ at which Q_J attains its maximum deviation from zero with alternating sign. It is apparent that these include the endpoints and the $J - 1$ minima and maxima of Q_J in $[a, b]$. When $t = J + 1$, the construction fails since $s = 2t - 1 - J = J + 1$ and the degree of P_s is greater than J.

Having shown that the function which has the least deviation from zero over $[a, b]$ alternates $J + 1$ times over the interval, we now note that any function with this property has the least possible deviation from zero. The de La Vallée Poussin theorem precludes the existence of two parameter sets, say $\mathbf{p}_m$ and $\mathbf{p}_k$, with this property and $H_m \ne H_k$.

Finally, we show that the parameter set which attains the least deviation is unique. We first note that $J + 1$ alternation points can be attained only when the J parameters are distinct. Also, $Q_J(x, \mathbf{p})$ is independent of the ordering of the parameters. We choose to express $\mathbf{p}$ as the J-tuple with elements ordered in increasing magnitude in the interval $[a, b]$. We now prove that the ordered solution to the minimax problem is unique. If two candidates have no common alternation points other than the endpoints, then the difference between the two functions vanishes at 0, a, and b and at $J - 2$ interior points which separate the interior $J - 1$ alternation points of one of the functions. This leads to the usual contradiction. The numerator in the difference function is of maximal degree $2J-1$ and has at least $2J+1$ roots. The only subtlety in the argument results from both functions having common extrema other than the endpoints. This can be resolved by careful counting of zeros of the difference function. At any common interior extremum, both the difference function and its derivative vanish. This requires at least a double root at each common extremum which may replace the two roots otherwise counted between that extremum and adjacent extrema.

1.3 Early Parameter Selection

A survey of early parameter choices with theoretical considerations and numerical comparisons may be found in Birkhoff, Varga and Young (1962). Although these parameters are adequate for most applications, there are more precise determinations which yield values as close to optimal as desired with little difficulty as well as associated bounds on error reduction. Peaceman and Rachford (1955) used $p_j = b(\frac{a}{b})^{(2j-1)/2J}$ and Wachspress (1957) used $p_j = b(\frac{a}{b})^{(j-1)/(J-1)}$. The algebra required to establish convergence rates and comparison with optimal parameters is tedious and will not be pursued here in view of the far more elegant theory now available. The logarithms of these initially used parameters are uniformly spaced.[2] When the parameters are spaced uniformly on a log scale, the extrema increase in magnitude toward the ends. One must shift the parameters somehow toward the ends to attain equal magnitudes for the extrema.

As already noted in Sect. 1.2, the first optimum parameter sets were developed independently for the case of $J = 2^n$ in 1962 by Gastinel and Wachspress. My analysis was motivated by discussions with Bengt Kredell, who during a visit to GE in Schenectady in 1961 expressed his disappointment that greater effort was not being expended in solving this intriguing minimax problem. This encouraged pursuit of a solution with renewed vigor. The character of this solution led to the more general result for all J demonstrated by W.B. Jordan [Wachspress, 1963]. Unfortunately, this theory was not applied to ADI iteration in time to be included in the thorough review in [Birkhoff et al., 1962] of the earlier ADI iteration theory through the analysis for $J = 2^n$.

[2] John Sheldon (personal communication) observed during the initial development period that the logarithms of the optimal parameters must be more closely spaced toward the ends of the interval. He drew the parallel with the Chebyshev parameter spacing for polynomial approximation. Sheldon's conclusion follows directly from the Chebyshev alternation property.

1.4 Optimum Parameters When $J = 2^n$

In the ensuing discussion, **p** is the unique solution to the minimax problem. The key to the solution for $J = 2^n$ was a simple observation which is the content of the following lemma:

Lemma 4 (Logarithmic Symmetry Lemma). *If* $p_j \in \mathbf{p}$, *then* $\frac{ab}{p_j} \in \mathbf{p}$.

Proof. Consider the factor in $Q_J(x, \mathbf{p})$ associated with parameter p_j:

$$F_j(x) = \frac{p_j - x}{p_j + x}. \tag{22.1}$$

Multiply numerator and denominator by $\frac{ab}{xp_j}$. Define $y = \frac{ab}{x}$ and $q_j = \frac{ab}{p_j}$. Then

$$-F_j(x) = G_j(y) = \frac{q_j - y}{q_j + y}. \tag{22.2}$$

The minus sign is irrelevant since we are concerned only with the absolute value of the product. Moreover, in this case, J is even and the minus sign drops out of the product. As x varies from a to b, y varies from b to a. Thus, the minimax problem is the same for $Q_J(x, \mathbf{p})$ and $Q_J(y, \mathbf{q})$. Hence, $\mathbf{p} = \mathbf{q}$.

Another obvious property of this minimax problem is that since $Q_J(x, \mathbf{p}) = Q_J(\sigma x, \sigma \mathbf{p})$ for any positive σ, the interval may be normalized to $[\frac{a}{b}, 1]$. If the optimum parameters for this interval are $\{p_j\}$, then the parameters for the actual interval are $\{bp_j\}$.

The log-symmetry lemma immediately establishes that when $J = 1$ the optimum single parameter is $p_1 = \sqrt{ab}$ and the corresponding value for H is $\frac{\sqrt{ab}-a}{\sqrt{ab}+a} = \frac{1-\sqrt{\frac{a}{b}}}{1+\sqrt{\frac{a}{b}}}$. This is a rather trivial result. The lemma may be used to a greater extent. Suppose J is even. Then one may combine the factors associated with p_j and $\frac{ab}{p_j}$ as follows:

$$\left[\frac{p_j - x}{p_j + x}\right]\left[\frac{\frac{ab}{p_j} - x}{\frac{ab}{p_j} + x}\right] = \frac{ab + x^2 - \left(p_j + \frac{ab}{p_j}\right)x}{ab + x^2 + \left(p_j + \frac{ab}{p_j}\right)x}. \tag{23}$$

Dividing numerator and denominator by $2x$, we express these factors as

$$\frac{\left[\frac{\frac{ab}{x}+x}{2}\right] - \left[\frac{\frac{ab}{p_j}+p_j}{2}\right]}{\left[\frac{\frac{ab}{x}+x}{2}\right] + \left[\frac{\frac{ab}{p_j}+p_j}{2}\right]}. \tag{24}$$

We now define

$$x^{(1)} \equiv \frac{\left(\frac{ab}{x} + x\right)}{2}, \tag{25.1}$$

$$p_j^{(1)} \equiv \frac{\left(\frac{ab}{p_j} + p_j\right)}{2} \tag{25.2}$$

for $j = 1, 2, \ldots, \frac{J}{2}$. For $x \in [a, b]$ we have $x^{(1)} \in [\sqrt{ab}, \frac{a+b}{2}]$. If the optimum parameters can be determined for the derived problem with $Q_{\frac{J}{2}}(x^{(1)}, \mathbf{p}^{(1)})$, then the optimum parameters for the initial problem can be obtained from Eq. 25.2. For each $p_j^{(1)}$, we obtain the two parameters

$$p_j = p_j^{(1)} + \sqrt{\left[p_j^{(1)}\right]^2 - ab}, \tag{26.1}$$

$$p_{1-j+J} = \frac{ab}{p_j}. \tag{26.2}$$

If $J/2$ is even the spectrum may be "folded" again to reduce the order of the problem to $J/4$. The solution $\mathbf{p}^{(2)}$ then yields $\mathbf{p}^{(1)}$ which in turn yields $\mathbf{p}$. When $J = 2^n$, the problem is first reduced to $J(n) = 1$ by n foldings. Then $p_1^{(n)}$ is just the square root of the interval endpoints after the foldings. Successive application of Eq. 26 yield $\mathbf{p}$. This is a "fan in–fan out" algorithm reminiscent of the fast Fourier transform algorithm.

The arithmetic–geometric mean (AGM) algorithm described in 22.20 in the NIST Handbook of Mathematical Functions plays a key role in this development. If the interval for $x^{(s)}$ is $[a^{(s)}, b^{(s)}]$, then $a^{(s)}$ is the geometric mean of $a^{(s-1)}$ and $b^{(s-1)}$ while $b^{(s)}$ is the arithmetic mean of $a^{(s-1)}$ and $b^{(s-1)}$. Index $s = 0$ is associated with the original interval and parameters. Thus, to compute the optimum parameters when $J = 2^n$, we first compute the successive intervals:

Interval AGM recursion formulas:

$$a^{(0)} = a \quad \text{and} \quad b^{(0)} = b,$$

$$a^{(1)} = \sqrt{a^{(0)}\, b^{(0)}} \quad \text{and} \quad b^{(1)} = \frac{a(0) + b(0)}{2},$$

$$a^{(s)} = \sqrt{a^{(s-1)}\, b^{(s-1)}} \quad \text{and} \quad b^{(s)} = \frac{a(s-1) + b(s-1)}{2}$$

for $s = 2, 3, \ldots, n$.

It is an easily established and well-known property of the AGM algorithm that the values form the nested sequence: $a^{(0)} < a^{(1)} < \cdots < a^{(n)} < b^{(n)} < b^{(n-1)} < \cdots < b^{(1)} < b^{(0)}$. The common limit as n increases is the "arithmetic–geometric mean" of $a^{(0)}$ and $b^{(0)}$.

The algorithm also yields a bound on the error reduction attained with the J ADI iterations. We note that since $J(n) = 1$,

$$H\left(\mathbf{p}^{(0)}\right) = H\left(\mathbf{p}^{(n)}\right) = \frac{\sqrt{a^{(n)}b^{(n)}} - a^{(n)}}{\sqrt{a^{(n)}b^{(n)}} + a^{(n)}} = \frac{a^{(n+1)} - a^{(n)}}{a^{(n+1)} + a^{(n)}}. \tag{27}$$

The bound on the reduction in the norm of the error is equal to H^2. Accurate formulas for relating error reduction to J and $\frac{a}{b}$ will be derived in the discussion of the solution to the minimax problem for all J. For the present, we assert that when $a << b$ and J is sufficiently large to yield an error reduction of $\rho(J) << 1$, a good approximation is

$$\rho(J) \doteq 4\exp\left[-\frac{\pi^2 J}{\ln \frac{4b}{a}}\right]. \tag{28}$$

The value of $\frac{b}{a}$ is a measure of the "condition" of the problem to be solved. The number of iterations for convergence to a prescribed accuracy varies for the model Peaceman–Rachford problem as $\ln \frac{b}{a}$. This compares quite favorably with $\sqrt{b/a}$ for successive overrelaxation and Chebyshev extrapolation and even with $[\frac{b}{a}]^{\frac{1}{4}}$ attainable with some of the more potent preconditioned conjugate gradient methods in common use. Unfortunately, the model-problem commutation condition is rather restrictive so that the logarithmic convergence rate can be assured only in special cases. Despite encouraging numerical results on an assortment of problems in which the commutation condition was not fulfilled, one should be reluctant to rely on a method lacking firm theoretical justification. For this reason, we concentrate here on applications with rigorous mathematical support.

1.5 Optimum Parameters from Chebyshev Alternance

A variety of minimax problems may be solved by application of the Chebyshev alternance property [Achieser, 1967]. This is illustrated by solution of the classical problem: Find the polynomial of maximal degree J, normalized to unity at $x = 0$ which deviates least from zero over the positive interval $[a, b]$. Although the solution is given in many texts, few describe how this result may be obtained directly from the Chebyshev alternance property. To solve this problem, one first establishes the Chebyshev alternance property, which is that the unique solution alternates $J + 1$ times over $[a, b]$. Let the maximum absolute value of this polynomial $P_J(x)$ be equal to the as yet unknown value H. Then, it is seen by comparing zeros of both sides that the following differential equation must be satisfied by $P_J(x)$:

$$J^2[H^2 - P_J^2(x)] = (x - a)(b - x)[P'(x)]^2. \tag{29}$$

Both sides are polynomials of degree $2J$ in x and have the same zeros and the same coefficient of x^{2J}. We take the square root of both sides, separate variables, and integrate from a to x, where x is less than its value at the first negative alternation point. A subtle point in the analysis is that P' is negative in this interval so that we obtain

$$\int_H^{P(x)} \frac{dP}{\sqrt{H^2 - P^2(x)}} = -J \int_a^x \frac{dx}{\sqrt{(x-a)(b-x)}}. \tag{30}$$

The substitutions $P(x) = H\cos\theta$ on the left and $x = \frac{1}{2}[(b+a) - (b-a)\cos\phi]$ on the right yield

$$P(x) = H\cos\left[J \arccos\left(\frac{b+a-2x}{b-a}\right)\right]. \tag{31}$$

That the right-hand side of Eq. 31 is a polynomial of degree J in x is easily proved with the three-term recursion formula based on the cosine identity: $\cos(n+1)w + \cos(n-1)w = 2\cos w \cos nw$. This must be true since the minimax problem is known to have a solution which satisfies the differential equation. We note that $\cos[J \arccos w] = \cosh[J \operatorname{arccos} h\, w]$ and that when $w < 1$ the cosine form is convenient while when $w > 1$ the hyperbolic cosine form is more convenient. The normalization to unity at $x = 0$ fixes H as

$$H = \frac{1}{\cosh J\left[\operatorname{arccos} h\left(\frac{b+a}{b-a}\right)\right]}. \tag{32}$$

This same technique may be used to solve the ADI minimax problem. In view of this classical approach to Chebyshev minimax problems, it is somewhat surprising that so many years elapsed between formulation (1957) and solution (1963) of the ADI minimax problem. The ADI problem is in fact identical to one solved by Zolotarev [Zolotarev, 1877] and the solution was used by Cauer in 1933 [Cauer, 1958] in design of electrical filters. A fascinating historical review and analysis of this minimax problem is given in [Todd, 1984].[3] The classical approach to the ADI minimax problem is to first note that $Q_J(-x) = 1/Q_J(x)$ and to construct the differential equation from the Chebyshev alternance property:

$$c[H^2 - Q(x)^2][H^{-2} - Q(x)^2] = (x^2 - a^2)(b^2 - x^2)[Q']^2, \tag{33}$$

where c is a constant determined by the condition $Q_J(0) = 1$. Since $Q_J(x)$ is the ratio of two monic polynomials of degree J, the numerators on both sides of Eq. 33 are polynomials of degree $4J$ and have $4J$ common roots, the simple roots

[3]One small point of which Todd was unaware is that Bill Jordan corresponded with Bode (whose *p–v* diagram is well known to engineers) on Cauer's analysis. It was my good fortune that Jordan joined our small math group at the Knolls Atomic Power Laboratory around the time I was working on this problem, for he provided the link with Zolotarev. Although I had read of this work in [Achieser, 1967] and suspected its relevance, I had not applied this theory to the ADI problem. I knew Bill was an expert on elliptic functions and welcomed his assistance.

at $x = a, -a, b, -b$, and double roots at the interior $J - 1$ positive extrema and $J - 1$ negative extrema. The common denominator in Eq. 33 is the fourth power of the denominator in $Q_J(x)$.

The solution is obtained by taking the square root, separating the variables Q and x, and evaluating the resulting elliptic integrals. Bill Jordan preferred a shortcut to this classical approach. Before proceeding with Jordan's analysis, we digress to state a few properties of elliptic functions. All the elliptic-function definitions and formulas needed here may be found in Chaps. 16 and 17 of [Abramowitz and Stegun, 1964] and in Chaps. 19 and 22 of [NIST Handbook of Mathematical Functions, 2010]. The ensuing discussion is best understood with one of these references in hand. The NIST handbook is online. References like "22.4.2" and "16.12" refer to tables and formulas in the these books.

The Jacobian elliptic functions of modulus k are defined "inversely" by the integral

$$z(\phi) = \int_0^\phi \frac{d\theta}{\sqrt{1 - k^2 \sin^2\theta}}. \tag{34}$$

The angle ϕ is called the amplitude and the three Jacobian elliptic functions are defined by $sn\, z = \sin\phi$, $cn\, z = \cos\phi$, and $dn\, z = \sqrt{1 - k^2 \sin^2\phi}$. The complete elliptic integral of the first kind is $K(k) = z(\frac{\pi}{2})$. The complementary modulus is $k' = \sqrt{1 - k^2}$. We define $K'(k) = K(k')$. These are the only doubly periodic meromorphic functions. The "quarter-periods" are defined as K and iK'. This definition is somewhat misleading since the periods of the three elliptic functions are not the same. The sn-function has real period $4K$ and imaginary period $2iK'$. The cn-function has real period $4K$ and imaginary period $4iK'$. The dn-function has real period $2K$ and imaginary period $4iK'$. The three functions all have poles at iK'. The "parameter" τ is defined as the ratio of the quarter periods: $\tau = \frac{iK'}{K}$ and the "nome" q is defined as $q = \exp[i\pi\tau] = \exp[-\pi\frac{K'}{K}]$. Two fundamental properties are that a rational function of elliptic functions is also an elliptic function (22.2) and if two elliptic functions have the same periods, poles, and zeros, then their ratio is a constant. Values of the dn-function which enter into Jordan's analysis include

$$dn\,(0) = 1, \quad dn\left(\frac{K}{2}\right) = \sqrt{k'}, \quad dn\,(K) = k',$$

$$dn\,(iK') = \infty, \quad \text{and } dn\,(K + iK') = 0. \tag{35}$$

The dn-function is an even function, $dn\,(-z) = dn\,(z)$, and it has odd symmetry about iK': $dn\,(z + 2iK') = -dn\,(z)$.

Various approximations and relationships among elliptic functions will be drawn upon as needed. These functions have very desirable numerical properties. Accurate evaluation is a simple task, and interrelationships of nomes, periods, and moduli facilitate analysis. Trigonometry is a basic secondary school subject. It is unfortunate that "elliptometry" is not a basic college undergraduate subject. Gauss played a significant role in the development of the theory of elliptic functions, and this theory is permeated with the elegance which is so characteristic of Gauss' work. Knowing that the solution to the ADI minimax problem was an elliptic function, Bill Jordan considered the normalized interval $[k', 1]$ and made the substitution $x = dn\,(uK, k)$. As x varies from k' to 1, u varies from 1 to 0. Then, $Q_J(x)$ is an elliptic function in u. We have

$$Q_J(u) = \prod_{j=1}^{J} \frac{p_j - dn\,(uK, k)}{p_j + dn\,(uK, k)}. \tag{36}$$

Now Bill jumped right to a form for Q_J which has the Chebyshev property:

$$Q_J(u) = (-1)^{J-1} \frac{\sqrt{k_1'} - dn\,(uJK_1, k_1)}{\sqrt{k_1'} + dn\,(uJK_1, k_1)}, \tag{37}$$

where k_1 is to be determined. This function alternates $J + 1$ times as u varies from 0 to 1 with extreme value $H = \frac{1-\sqrt{k_1'}}{1+\sqrt{k_1'}}$. We must choose k_1 so that the right-hand sides of Eqs. 36 and 37 have the same periods and poles. We then choose the p_j so that they have the same zeros. Finally, we show that their constant ratio is equal to unity.

For any choice of the p_j, the right-hand side of Eq. 36 has real period 2 and the right-hand side of Eq. 37 has real period $2/J$. Since a function with period $2/J$ also has period 2, both functions have the same real period of 2. (Only when the p_j are chosen to equate these two functions will the right-hand side in Eq. 36 have real period $2/J$ as well.) For any p_j, the right-hand side of Eq. 36 has parameter $\tau \equiv iK'/K$ and imaginary period $4iK'$. Similarly, the right-hand side of Eq. 37 has parameter τ_1/J. The imaginary periods are equal if we choose $\tau_1 = J\tau$. There is a $1 - 1$ correspondence between modulus k and parameter τ so that this equating of imaginary periods fixes k_1.

Having equated periods for any choice of the p_j, we now choose them to equate zeros. The zeros in Eq. 37 are at $u = \frac{2j-1}{2J}$ for $j = 1, 2, \ldots, J$. (The value of the dn-function is equal to the square root of its complementary modulus at odd multiples of K_1.) We therefore choose $p_j = dn\,[\frac{(2j-1)K}{2J}, k]$. It remains to be shown that the two functions then have the same poles. The poles occur when the dn-functions are the negatives of their values at the roots. The arguments must therefore be increased by $2iK'$. Thus, if u_j is a root, then $u_j + 2\tau$ is a pole in Eq. 36. Plugging this into Eq. 37, we find that

$$\begin{aligned} dn\,[(u_j + 2\tau)JK_1, k_1] &= dn\left[\frac{(2j-1)K_1}{2} + 2\tau_1 K_1, k_1\right] \\ &= dn\left[\frac{(2j-1)K_1}{2} + 2iK_1', k_1\right] \\ &= -dn\left[\frac{(2j-1)K_1}{2}, k_1\right] = -\sqrt{k_1'}. \end{aligned}$$

We have not examined all roots and poles in the complete "period rectangle" of the elliptic function, which extends from $u = 0$ to $u = 2$ along the real axis and from $u = 0$ to 4τ along the imaginary axis. That the roots and poles are identical for the complete rectangle is readily deduced from the elliptic function identity (16.8.3): $dn\,(z + K, k) = k'/dn\,(z, k)$. The roots and poles are thus symmetric in the first period rectangle with respect to the line $u = 1$.

To show that the ratio of the functions in Eqs. 36 and 37 is unity, we choose $u = \tau$. Then the dn-functions become infinite (16.5.7) in both equations, and both functions are equal to $(-1)^J$. This completes Jordan's proof that the optimum ADI parameters for the eigenvalue interval $[k', 1]$ are

$$p_j = dn\left[\frac{(2j-1)K}{2J}, k\right], \quad j = 1, 2, \ldots, J. \tag{38}$$

We can play the "matching game" one more time to prove that

$$Q_J(u) = (-1)^J \sqrt{k_2}\, sn\,[(1 + 2uJ)K_2, k_2], \tag{39}$$

where $\tau_2 = 4\tau_1$. We compare the right-hand side of Eq. 39 with the right-hand side of Eq. 37. The zeros of both functions are at uJ an odd integer. The real period of the sn-function is $4K$ (16.2) and of the dn-function is $2K$ so that both functions have the same real period with respect to u of $\frac{2}{J}$. The imaginary period of the dn-function is $4iK'$ (16.2) and of the sn-function is $2iK'$. Thus, the imaginary periods are the same when $J\frac{K_2'}{K_2} = 4J\frac{K_1'}{K_1}$ or when $\tau_2 = 4\tau_1$. The poles of the sn-function (16.2) are at the zeros plus iK', and hence both functions have the same poles. It remains to be proved that the two functions are equal at some value of u. At $u = \tau$, we have already shown that the right-hand side of Eq. 37 is $(-1)^J$. Since $\tau = \tau_2/4J$, the argument of the sn-function in Eq. 39 is $K_2 + 2uJK_2 = K_2 + \frac{\tau_2 K_2}{2} = K_2 + \frac{iK_2'}{2}$, and by 16.8 and 16.5, $sn\,(K + \frac{iK'}{2}) = \frac{cn\,\frac{iK'}{2}}{dn\,\frac{iK'}{2}} = k_2^{-\frac{1}{2}}$. Thus, the right-hand side of Eq. 39 is equal to $(-1)^J$ when $u = \tau$ and the identity is established.

Elliptic function identities in which the parameter is increased (decreased) by a factor of two are "ascending" ("descending") Landen or Gauss transformations (N-22.7). The equality of the right-hand sides of Eqs. 37 and 39 can be established directly through inverse application of the descending Landen transformation to relate $dn\,(z, k_1)$ to $dn^2(v, k_3)$ and then application of the ratio of the ascending Landen transformations to relate $dn^2(v, k_3)$ to $\frac{cn\,(w,k_2)}{dn\,(w,k_2)}$. The latter is equal to $sn\,(w + K_2, k_2)$. In the above, k_3 is an intermediate modulus related to k_1 through 16.12.1, and the relationships between u, v, and w are given in 16.12.1 and 16.14.1 along with the relationship of k_3 to k_2. Cauer's solution to an equivalent minimax problem was expressed in terms of the sn-function. Cauer solved an electrical filter design problem which led to the minimax problem:

$$\min_{\{a\}} \max_{0 \le z \le h < 1} \left| \prod_{j=1}^{t} \frac{a_j^2 - z^2}{1 - a_j^2 z^2} \right| \tag{40}$$

for a_j real. As shown on pp. 192–193 in Wachspress (1966), this may be transformed to the ADI minimax problem by the birational change of variables from z to x:

$$z^2 = \frac{1-x}{1+x}. \tag{41}$$

The solution of the ADI minimax problem involved a modular transformation of an elliptic function which increased the parameter by a factor of J. An increasing Landen transformation doubles the parameter. The algorithm for determining the optimum parameters when $J = 2^n$ was actually a succession of Landen transformations in which the interval bounds were related by geometric and arithmetic means. The AGM algorithm plays a crucial role in the theory and evaluation of elliptic functions and elliptic integrals.

1.6 Evaluating the Error Reduction

Referring to Eq. 39, we observe that the ADI error reduction after J iterations with the optimum parameters is

$$R_J = H_J^2 = \max_{k' \le x \le 1} Q_J^2(x, \mathbf{p}) = k_2, \tag{42}$$

where k_2 is the modulus of the elliptic function whose nome is

$$q_2 = q(k)^{4J} = \exp\left[-\frac{4\pi J K'}{K}\right]. \tag{43}$$

Here K and K' are the elliptic integrals associated with the complementary modulus $k' = \frac{a}{b}$. From the known value for k', one may use the AGM algorithm to compute K and K' to any desired accuracy. For a prescribed error reduction ε and eigenvalue interval $[a, b]$, we must first compute the smallest J for which $k_2 < \varepsilon$ and then the values for the associated p_j. Various approximations have been used. The computation of values to eight significant digits is a relatively simple task. Earlier applications of ADI iteration were for both k' and ε much less than unity. More recent applications lead to values which can approach unity [Wachspress, 1990]. Sometimes the value of J is given and the associated error bound is to be determined.

Given any one of k, k', q, q', the other three values may be computed readily. We first note that

$$k^2 + k'^2 = 1 \tag{44.1}$$

Table 1.1 An AGM table

n	a_n	b_n	c_n	ϕ_n
0	1	k'	k	ϕ_0
1	$\frac{1}{2}(a_0+b_0)$	$\sqrt{a_0b_0}$	$\frac{1}{2}(a_0-b_0)$	ϕ_1
2	$\frac{1}{2}(a_1+b_1)$	$\sqrt{a_1b_1}$	$\frac{1}{2}(a_1-b_1)$	ϕ_2
...	...	...	...	...
N	$\frac{1}{2}(a_{N-1}+b_{N-1})$	$\sqrt{a_{N-1}b_{N-1}}$	$\frac{1}{2}(a_{N-1}-b_{N-1})$	ϕ_N

and

$$\ln q \ln q' = \pi^2 \tag{44.2}$$

relate values of modulus and nome to complementary modulus and nome.

When q (or q') $< 10^{-4}$, k(or k') may be computed with eight-digit accuracy with the approximation

$$k \doteq \frac{4q^{\frac{1}{2}}}{(1+2q)^2} \tag{45.1}$$

or

$$k' \doteq \frac{4q'^{\frac{1}{2}}}{(1+2q')^2}. \tag{45.2}$$

These formulas give four-digit accuracy when q or $q' < 0.01$. Approximate reversion of these formulas yields

$$q \doteq \frac{k^2}{16}\left(1+\frac{k^2}{4}\right)^2, \tag{45.3}$$

$$q' \doteq \frac{k'^2}{16}\left(1+\frac{k'^2}{4}\right)^2. \tag{45.4}$$

These give four-digit accuracy when k or k' is less than 0.1 and seven-digit accuracy when less than 0.04. Before giving formulas relating nomes to moduli when Eq. 45 are insufficient, we describe how values may be computed to any desired accuracy in general.

For modulus k, we construct an AGM table. Values for a_n, b_n, and c_n are computed recursively from the first row down. Values for ϕ_n are computed in the same order when evaluating the elliptic integral for argument ϕ_0. The recursion formulas for ϕ will be described subsequently. This computation is not needed when dealing with real eigenvalue intervals but is needed when the theory is generalized into the complex plane. Values for ϕ_n are computed in reverse order when evaluating elliptic functions (Table 1.1).

The algorithm is terminated when $\frac{c_N}{a_N}$ is less than a prescribed tolerance. The AGM of 1 and k' is then approximated by a_N. The complete elliptic integral of the first kind is $K(k) = \frac{\pi}{2a_N}$. The complementary integral K' is obtained by repeating the AGM algorithm with $b_0 = k$ and $c_0 = k'$. If this AGM is a_M, then $K'(k) = \frac{\pi}{2a_M}$. This procedure is consistent with the identity $K'(k) = K(k')$.

The Jacobi elliptic functions of argument u and modulus k are computed by backward recursion from ϕ_N to ϕ_0, where $\phi_N = 2^N a_N u$ and

$$\phi_{n-1} = \frac{1}{2}\left[\phi_n + \arcsin\left(\frac{c_n}{a_n}\sin\phi_n\right)\right], \tag{46}$$

where the arcsin is bounded in absolute value by $\frac{\pi}{2}$. The values for the three Jacobi elliptic functions are computed as $sn\,u = \sin\phi_0$, $cn\,u = \cos\phi_0$, and $dn\,u = \frac{\cos\phi_0}{\cos(\phi_1-\phi_0)}$. For values of u close to K, the latter formula is sensitive to roundoff and it is better to compute $sn\,u$ and then $dn\,u = \sqrt{1-k^2\,sn^2\,u}$. Periodic properties of the elliptic functions are such that one may always express an elliptic function of any real argument in terms of elliptic functions with arguments in the interval $[0, K]$ (Table 16.8).

The J values for u needed for determining the p_j are $\frac{(2j-1)K}{2J}$. Since $a_N = \frac{\pi}{2K}$, the appropriate values for ϕ_N are $\phi_N(j) = \frac{2^{N-2}\pi(2j-1)}{J}$.

Once $q(k)$ is computed, the next step depends on whether J or ε is given. Suppose the latter error reduction is prescribed. Then we know that $k_2 \le \varepsilon$. In practice J must be an integer and we choose the smallest J for which $k_2 < \varepsilon$. The AGM algorithms for modulus k_2 yield K_2, K_2', and hence q_2. The value for J is

$$J = \left[\left[\frac{\ln q_2}{4\ln q}\right]\right], \tag{47}$$

where the symbol $[[x]]$ denotes "the smallest integer greater than x." Having found J in this fashion or when given J, we may wish to compute the corresponding value for $k_2 = \varepsilon$. We first compute $q_2 = q^{4J}$ and next determine the associated value for k_2. If $q_2 \ge \exp(-\pi)$, we first compute $q_2' \le [\exp(-\pi) = 0.043214]$ from

$$q_2' = \exp\left[\frac{\pi^2}{\ln q_2}\right]. \tag{48}$$

If either q_2 or ${q'}_2 < 10^{-4}$, Eqs. 44–45 may be used to obtain k_2. Otherwise, the basis for approximating k_2 is the ratio of 16.38.8 to 16.38.5, which when $q' < 0.0433$ yields at least eight-digit accuracy when truncated to

$$k \doteq \left[\frac{1-2q'+2q'^4}{1+2q'+2q'^4}\right]^2. \tag{49}$$

When $q_2' = e^{-\pi}$, this yields $k_2 \doteq 0.70710678121$ on my electronic slide rule as compared with truth equal to $1/\sqrt{2} = 0.707106781184$. The formula is more accurate when q' is smaller. If $q_2 < e^{-\pi}$, we substitute q_2 for q' in Eq. 49 to compute

Table 1.2 An AGM table

n	a	b	c	$\phi(1)$	$\phi(2)$	$\phi(3)$	$\phi(4)$
0	1.0000	0.0100	0.99995	0.6874	1.3600	1.5240	1.5626
1	0.5050	0.1000	0.4950	0.6956	1.4070	1.7346	2.4460
2	0.3025	0.2247	0.2025	0.8594	2.2820	4.0010	5.4237
3	0.2636	0.2607	0.0389	$\pi/2$	$3\pi/2$	$5\pi/2$	$7\pi/2$
4	0.2622	0.2622	0.0014	π	3π	5π	7π

k_2' to high accuracy and then $k_2 = \sqrt{1-(k_2')^2}$. Four-digit accuracy is maintained when the fourth powers are omitted on the right-hand side of Eq. 49. Reversion of this approximation yields q and q' as a function of k or k'. Suppose, for example, $k > 1/\sqrt{2}$. Then if we define

$$z \equiv \frac{q'}{1+q'^4}, \tag{50.1}$$

we find that

$$z \doteq \frac{1}{2}\frac{1-\sqrt{k}}{1+\sqrt{k}}. \tag{50.2}$$

is a good approximation to q' and that an even better approximation is

$$q' \doteq z(1+z^4). \tag{50.3}$$

For example, when $k = 1/\sqrt{2}$, $z = 0.043213616$ and Eq. 50.3 give the value of 0.043213918, which is correct to all digits displayed. Equations 49–50 can suffer from roundoff when any of the moduli or nomes is very small, and for this reason the approximations in Eqs. 44–45 should be used when applicable. For example, to compute error reduction when $k' = 1/\sqrt{2}$ and $J = 2$, we find that $q_2 = \exp(-8\pi) = 1.216 \times 10^{-11}$ and Eqs. 49–50 yield $k_2' = 1.0\ldots$ to nine-digit accuracy so that $k_2 = 0$. However, Eq. 45 yields $k_2 \doteq 4\exp(-4\pi) = 1.3949\times 10^{-5}$, which is correct to the five displayed digits (Table 1.2).

The AGM algorithm will now be illustrated for the case where $k' = 0.01$ and $J = 4$.

We compute $K = \frac{\pi}{2(0.2622)} = 5.991$. The AGM for 1 and k yields $K' = 1.5708$. We compute $q' = \exp[-\frac{\pi K}{K'}] = 6.25 \times 10^{-6}$. The approximation in Eq. 50.3 gives the same value. The values for the dn-function obtained from the above table are $dn\,\frac{K}{8} = 0.7729$, $dn\,\frac{3K}{8} = 0.2095$, $dn\,\frac{5K}{8} = 0.0478$, and $dn\,\frac{7K}{8} = 0.01292$. These values agree with those obtained by the square-root algorithm described in Sect. 1.4.

The associated error reduction is readily obtained. We compute $q = \exp[\frac{\pi^2}{\ln q'}] = 0.43883$ and $q_2 = q^{16} = 1.89 \times 10^{-6}$. Now Eq. 49 yields $k_2' = 0.99998488$ and $k_2 = 0.005499$, which agrees with the value determined by the square-root algorithm: $\varepsilon = \left(\frac{b_3-b_2}{b_3+b_2}\right)^2$. We note that Eq. 45.1 can be used here to obtain $k_2 = 0.005499$.

Error reduction estimates for applications involving complex spectra require evaluation of incomplete elliptic integrals. These are computed with the aid of the AGM algorithm. Now u is the integral $F(\phi, k)$ and the first entry in the last column of the AGM table is $\phi_0 = \phi$. The angles in the last column are now computed with forward recursion formulas which correspond to descending Landen transformations. These are given in 17.6.8 as

$$\phi_n = \phi_{n-1} + \arctan\left(\frac{b_{n-1}}{a_{n-1}} \tan\phi_{n-1}\right). \tag{51}$$

(This equation was given incorrectly in [Wachspress, 1990]. The values for b and a were one row off in that reference. This led to slightly incorrect values for error reductions approximated there with Eq. 3.29.)

Although not stated in Abramowitz and Stegun but observed in the NIST handbook, it is crucial that the correct arctan be used in Eq. 51, and this is the angle in the interval $[\phi_{n-1} - \frac{\pi}{2}, \phi_{n-1} + \frac{\pi}{2}]$. The value of the incomplete elliptic integral of the first kind is $u = \frac{\phi_N}{2^N a_N}$. Use of Eq. 51 in computation of error bounds over complex spectra will be deferred until after the theory for ADI iteration with complex spectra is exposed.

1.7 Approximate Parameters

1.7.1 Error Reduction

For $J = 1$ or 2, the AGM algorithm in Sect. 1.4 yields the error reduction ε_J with little effort:

$$\varepsilon_1 = \left[\frac{1 - \sqrt{k'}}{1 + \sqrt{k'}}\right]^2, \tag{52.1}$$

and

$$\varepsilon_2 = \left[\frac{1 - \sqrt{k_1'}}{1 + \sqrt{k_1'}}\right]^2 \tag{52.2}$$

where

$$k_1' = \frac{2\sqrt{k'}}{1 + k'}. \tag{52.3}$$

When $k' < 0.1$, a reasonable approximation is

$$q \doteq \exp\left[-\frac{\pi^2}{2\ln(4/k')}\right], \tag{53}$$

and when $q^J < 0.3$ (in which case $\varepsilon_J < 0.36$):

Table 1.3 ADI parameter approximation

j	1	2	3	4
Truth	0.9873	0.8985	0.7870	0.7162
Eq. 56	0.9869	0.8984	0.7870	0.7165
Eq. 57	0.9146	0.8815	0.8022	0.7731

$$\varepsilon_J = k_2(J) = 4q^{2J} \doteq 4\exp\left[-\frac{\pi^2 J}{\ln(4/k')}\right]. \tag{54}$$

Combining these approximations, we may approximate J by

$$J \doteq \left[\left[\frac{1}{\pi^2}\ln(4/k')\ln(4/\varepsilon)\right]\right]. \tag{55}$$

If J computed with this last equation is less than 3, we may apply Eq. 52 to verify ε.

Equations 54–55 are adequate for most problems with real spectra that occur in practice. When $k' > 0.1$ or when higher accuracy is demanded we may use the simple relationships in Eqs. 44–50.

1.7.2 Iteration Parameters

It is almost always true in applications with real spectra that the iteration parameters may be computed with an approximate formula that has negligible adverse effect on error reduction. A modified version of the formula given by Jordan on p. 191 in [Wachspress, 1966] is recommended:

$$dn\,(rK,k) \doteq w(r) = \sqrt{k'}q'^{\frac{2r-1}{4}}\frac{1+q'^{1-r}+q'^{1+r}}{1+q'^{r}+q'^{2-r}}. \tag{56}$$

This approximation preserves the log-symmetry in that $w(r)w(1-r) = k'$. The iteration parameters are $w_j = w(r_j)$ for $j = 1, 2, \ldots, J$, where $r_j = (2j-1)/2J$. An even simpler approximation given on p. 191 of [Wachspress, 1966] is

$$w(r) = 2(k'/4)^r\frac{1+(k'/4)^{2(1-r)}}{1+(k'/4)^{2r}}. \tag{57}$$

This last formula is quite accurate when $k' < 0.01$. However, for larger values of k', one should use Eq. 56. Values are tabulated below for $k' = 1/\sqrt{2}$ and $J = 4$ (Table 1.3).

When $k' = 0.01$ and $J = 4$, we find that the three methods agree to four significant digits.

Table 1.4 Spectral radii comparison

k'	J =	Spectral 4	Radius 8	
10^{-2}		0.011	4.25×10^{-5}	Wachspress
10^{-2}		0.0055	7.57×10^{-6}	Optimum
10^{-4}		0.1668	4.76×10^{-3}	Wachspress
10^{-4}		0.0964	2.32×10^{-3}	Optimum

1.8 Use of Other Data on Spectra and Initial Error

One should bear in mind the analysis in Guilinger (1965) and in [Lynch and Rice, 1968] relating to choice of parameters when the initial error is known to be smooth. The ADI minimax function may then be modified by a positive weighting factor which accounts for an estimated spectral density of the error. For example, the low end of the spectrum may be more heavily weighted. Even in such cases, it is recommended that after an initial cycle, one should revert to the usual parameters. Spectral resolution of the initial error is rare in practice. Optimum parameters in such rare cases may be found by a Remez-type algorithm, requiring more than a little effort in implementation. Lynch and Rice recommend use of the "Wachspress" parameters because the interval bounds are two of the parameters and these act on denser eigenvalues near these endpoints more efficiently than the dn-parameters which yield the equal extreme property associated with the solution to the Chebyshev minimax problem. However, a more careful consideration of these parameters shows that the maxima occur near the endpoints and are larger in magnitude than the spectral radius with the elliptic-function parameters (Table 1.4). For example:

Convergence analysis for the "Wachspress" parameters was first given in [Wachspress and Habetler, 1960] and subsequently in [Birkhoff, Varga and Young, 1962].

The optimum parameters determined by Lynch and Rice for particular spectral density functions are more closely packed near the low end of the spectrum and thus yield greater reduction of these low error modes. My inclination is to use the elliptic parameters for arbitrary initial error and be assured of the associated error reduction. When there is a known bias toward the low end, one might modify this strategy by performing a few initial iterations with $p_j = k'$. A difficulty with strategies which attempt to account for smoothness of the initial error is uncertainty in accuracy achieved. Residual norms attach low weights to the low end of the spectrum and may be misleading.

Analysis of the Laplacian operator over a square grid of 100×100 points is revealing. The eigenvalues of H and V are

$$\lambda_m = 2\left(1 - \cos\frac{m\pi}{100}\right),$$
$$\gamma_n = 2\left(1 - \cos\frac{n\pi}{100}\right).$$

The eigenvalue interval for H and V is $[a, b] \doteq [0.000987, 4]$. The number of iterations for error reduction ε with the dn-parameters is

$$J_0(\varepsilon) \doteq \frac{2}{\pi^2} \ln \frac{4}{\varepsilon} \ln \frac{400}{\pi}.$$

When $\varepsilon = 10^{-6}$, $J_0 = 15$. If the $k-1$ smallest eigenvalues are eliminated by preliminary iteration with parameters equal to these eigenvalues, then the total number of iterations is $J_{k-1} = [k - 1 + 3.08 \ln \frac{400}{k\pi}]$. This is minimized when $k = 3$, in which case $J_2 = 14$. This savings of one iteration can hardly be considered significant. If, on the other hand, an error reduction of $\varepsilon = 10^{-9}$ is demanded, then $J_0 = 22$, $k = 4$ is best, and $J_3 = 19$. This could be worth the effort. There is probably a greater saving if the lower modes have more significant components in an expansion of the initial error.

When the spectrum is split, it may be even more beneficial to adapt parameters to the known split spectrum. For example, if the eigenvalues fall in $[10^{-6}, 10^{-4}] \cup [0.01, 1.]$ and $\varepsilon = 10^{-6}$, $J_0 = 24$ for $[a, b] = [10^{-6}, 1]$ while if one reduces the error for each interval separately, $J = 20$. If $\varepsilon = 10^{-9}$ with this split spectrum, $J_0 = 34$ and $J = 28$. When we deal with complex spectra, it will be shown that great improvement is sometimes possible through use of parameters which concentrate on isolated spectral regions.

Having analyzed the PR ADI iteration of Eq. 2, we now turn our attention to the generalized ADI iteration of Eq. 3. This is a rational approximation to zero in two independent variables with $2J$ free parameters. It is rare that an analytic solution can be found for problems of this complexity, and we are indeed fortunate that such a solution has been found for this particular problem.

Chapter 2
The Two-Variable ADI Problem

Abstract When the eigenvalue intervals for the commuting ADI matrices are not the same, the iteration is generalized by allowing different parameters for the two sweeps of each iteration. William B. Jordan demonstrated how one may reduce the two-variable minimax problem to one variable and obtain optimal parameters for the two sweeps. I subsequently resolved a basic assumption in his analysis in my PhD thesis which is summarized here. Application to three space dimensions is considered. A brief discussion of a different number of sweeps in each two-step iteration is also given.

2.1 Rectangular Spectra

We have developed a satisfactory theory for the Peaceman–Rachford ADI iterative solution of model problems where matrices H and V have the same spectral intervals. That improvement is possible when these intervals differ is demonstrated with a simple example. Let the interval for H be $[0.001, 4]$ and for V be $[0.025, 4]$. A prescribed error reduction yields a value for the nome q_2. Referring to Eqs. 1–43, we find that the number of iterations varies as K/K'. When $k' << 1$ this varies as $\ln \frac{4}{k'}$. For straightforward use of Eq. 2 of Chap. 1, we would choose parameters for the eigenvalue interval $[0.001, 4]$, and the number of iterations would be $J \doteq s \ln \frac{4}{k'} = s \ln \frac{16}{0.001} = 9.68s$ for some constant s depending on the prescribed error reduction. Suppose we redefine H and V by adding $\frac{c-a}{2} = 0.012$ times the identity matrix to H and subtracting this from V. The new eigenvalue intervals are $[0.013, 4.012]$ and $[0.013, 3.988]$. We find that for these intervals $J \doteq s \ln \frac{16.048}{0.013} = 7.12s$, and we have a significant gain in efficiency.

Inspection of Eq. 2 of Chap. 1 reveals that this is equivalent to retaining the original H and V matrices but using different iteration parameters in Eqs. 1–2.1 and 1–2.2. If the parameters for the redefined matrices are p_j, then we could use $p'_j = p_j + 0.012$ with the original H in Eq. 1–2.1 and $q'_j = p_j - 0.012$ with the original V in Eq. 1–2.2. We, therefore, generalize the Peaceman–Rachford equations

E. Wachspress, *The ADI Model Problem*, DOI 10.1007/978-1-4614-5122-8_2,
© Springer Science+Business Media New York 2013

to Eq. 3 of Chap. 1 (with matrix F equal to the identity for the present). One now considers optimization of these generalized equations. In our illustrative example, the simple shift led to almost identical eigenvalue ranges, and little gain could be achieved by further optimization. However, suppose the intervals for the eigenvalues of H and V were $[0.01, 1]$ and $[1, 100]$. The shift to equate lower bounds at 0.505 leads to upper bounds of 1.495 and 100.495. This gives a partial improvement from $k' = 0.0001$ to $k' = 0.005$, but greater improvement is possible. Before describing how this is accomplished, we consider the ADI minimax problem for Eqs. 3 of Chap. 1. The spectral radius of the generalized ADI iteration (GADI) matrix is

$$\rho(G_J) = \max_{\lambda,\gamma} \left| \prod_{j=1}^{J} \frac{(q_j - \lambda)(p_j - \gamma)}{(p_j + \lambda)(q_j + \gamma)} \right|,$$

where λ ranges over the eigenvalues of $F^{-1}H$ and γ ranges over the eigenvalues of $F^{-1}V$.

When F is the identity matrix, this is the 2-norm of the ADI iteration matrix. The B-norm of a vector $\mathbf{v}$ for any SPD matrix B is defined as the square root of the inner product $(\mathbf{v}, B\mathbf{v})$. The subordinate matrix norm is called the B-norm of the matrix. In general, the spectral radius of the ADI iteration matrix G_J is equal to the F-norm of G_J, and we choose to define our minimax problem as minimization of this norm. This norm is equal to the 2-norm of $F^{\frac{1}{2}} G_J F^{-\frac{1}{2}}$ which is equal to $\rho(G_J)$. Thus, the minimax problem for the generalized ADI equations is for a given J to choose sets of iteration parameters p_j and q_j to minimize $\rho(G_J)$. The role of matrix F will be developed later. For the present, we choose F as the identity matrix. Suppose λ and γ both vary over the same interval. Then we may revert to Eq. 2 of Chap. 1 by choosing $p_j = q_j$. It happens that this choice is optimal. Although this seems evident from symmetry considerations with respect to the two eigenvalue variables, the proof is not trivial and will be given subsequently. It follows that the additional degrees of freedom in Eq. 3 of Chap. 1 lead to a more efficient scheme only when the eigenvalue intervals differ.

2.2 W.B. Jordan's Transformation

The algorithm for $J = 2^n$ was generalized by Jordan to yield optimum parameters when the eigenvalue intervals were $[a, b]$ and $[c, d]$ with $a + c > 0$ [Wachspress, 1963, 1966]. Before each reduction of order (fan-in) by spectrum folding, the spectra were shifted by adding a constant to one and subtracting that constant from the other so that the product of the endpoints was identical for the shifted spectra. This enabled a folding that preserved the original form of the error function. Significant improvement was demonstrated when these intervals differed widely. Just as the earlier algorithm with its AGM theme stimulated W.B. Jordan to develop the elliptic-function theory for all J for Eq. 2 of Chap. 1, the generalized algorithm for Eq. 3 of Chap. 1 led Jordan to solution of this minimax problem. He found a transformation of variables which preserved the form of the minimax problem but with identical ranges for the new variables [Wachspress, 1963, 1966].[1]

A linear fractional transformation $y = B(z)$ is of the form

$$y = \frac{\alpha z + \beta}{\gamma z + \delta}. \tag{1}$$

The composite transformation $B(z) = B_2[B_1(z)]$ is isomorphic to matrix multiplication with

$$B \sim \begin{bmatrix} \alpha & \beta \\ \gamma & \delta \end{bmatrix}. \tag{2}$$

Thus, the composite transformation is obtained with $B = B_2 B_1$. Moreover, if we define $B_-(z) = B(-z)$, then

$$B_- = B \begin{bmatrix} -1 & 0 \\ 0 & 1 \end{bmatrix}. \tag{3}$$

The two-variable ADI minimax problem is to find the parameters p_j and q_j which minimize the maximum absolute value of the function

$$g(x, y, \mathbf{p}, \mathbf{q}) = \prod_{j=1}^{J} \frac{(x - q_j)(y - p_j)}{(x + p_j)(y + q_j)} \tag{4}$$

for $x \in [a, b]$ and $y \in [c, d]$, where $a + c > 0$. Define the linear fractional transformation

$$R_j(z) = \frac{z - q_j}{z + p_j}. \quad \text{Then} \quad \frac{(x - q_j)(y - p_j)}{(x + p_j)(y + q_j)} = \frac{R_j(x)}{R_j(-y)}.$$

[1]The development of this theory was exciting for both Bill and me, and our office-mates had to endure animated discussions between us over a period of several days. They were spared the nightly phone calls as we pursued this after hours. I recall the morning when Bill arrived for work with the solution in head. He approached the blackboard, rolled up his shirtsleeves with the comment "nothing up this sleeve" with each sleeve. In retrospect, Bill always felt that this transformation was the most elegant part of the analysis. After all, the elliptic-function theory had been developed 100 years earlier and had only to be introduced for this application. Bill's original analysis utilized relationships that were clear to Bill but obscure to me, and his derivation has to my knowledge never been published. I devoted significant effort to devising an alternative exposition and have found nothing more satisfactory than the approach resting on an isomorphism between linear fractional transformations and order-two matrix algebra which will now be presented.

We now seek a relationship between transformations B_1 and B_2 such that when we define $x = B_1(x')$ and $y = B_2(y')$ there exist a $\mathbf{p}'$, $\mathbf{q}'$ such that $g(x, y, \mathbf{p}, \mathbf{q}) = g(x', y', \mathbf{p}', \mathbf{q}')$. This can be accomplished if for each j

$$\frac{R_j(x)}{R_j(-y)} = \frac{R_j[B_1(x')]}{R_j[-B_2(y')]} = \frac{S_j(x')}{S_j(-y')} \tag{5}$$

for some linear fractional transformation S_j.

The matrix isomorphism yields $R_j B_1 = S_j$ and $R_{j-} B_2 = S_{j-}$. Thus, $R_j B_1 = (R_{j-} B_2)_-$, and it follows that

$$R_j B_1 = R_j \begin{bmatrix} -1 & 0 \\ 0 & 1 \end{bmatrix} B_2 \begin{bmatrix} -1 & 0 \\ 0 & 1 \end{bmatrix}, \tag{6}$$

and multiplying on the left by R_j^{-1}, we find that our goal is achieved when

$$B_1 = \begin{bmatrix} -1 & 0 \\ 0 & 1 \end{bmatrix} B_2 \begin{bmatrix} -1 & 0 \\ 0 & 1 \end{bmatrix}. \tag{7}$$

This yields the desired relationship between B_1 and B_2:

$$\text{If } B_1 = \begin{bmatrix} \alpha & \beta \\ \gamma & \delta \end{bmatrix}, \text{ then } B_2 = \begin{bmatrix} \alpha & -\beta \\ -\gamma & \delta \end{bmatrix}.$$

In Chap. 1 it was demonstrated that the optimum parameters for the one-variable problem with $x \in [a, b]$ are $p_j = q_j = \mathrm{b}dn[\frac{(2j-1)K}{2J}, k]$, where $dn[z, k]$ is the Jacobian elliptic dn-function of argument z and modulus $k = \sqrt{1 - k'^2}$. Here, k' is the complementary modulus, which is in this application equal to $\frac{a}{b}$. Having this result in mind, Jordan chose to normalize the common interval of x' and y' to $[k', 1]$. We now derive Jordan's result, which is that there is a unique $k' < 1$ and transformation matrix B_1 which accomplishes this task. The four conditions, when $x = a,\ x' = k'$; when $x = b,\ x' = 1$; when $y = c,\ y' = k'$; and when $y = d,\ y' = 1$, yield the homogeneous matrix equation $C\boldsymbol{\phi} = \mathbf{0}$, where $\boldsymbol{\phi}^T = [\alpha, \gamma, \beta, \delta]$ and

$$C = \begin{bmatrix} k' & -ak' & 1 & -a \\ 1 & -b & 1 & -b \\ k' & ck' & -1 & -c \\ 1 & d & -1 & -d \end{bmatrix}. \tag{8}$$

This system has a nontrivial solution only when the determinant of matrix C vanishes. It will be shown that there are only two values for k' for which this occurs, one greater than unity and the other less than unity. We first define the three matrices:

$$K = \begin{bmatrix} k' & 0 \\ 0 & 1 \end{bmatrix}, \quad A = \begin{bmatrix} 1 & -a \\ 1 & -b \end{bmatrix}, \text{ and } F = \begin{bmatrix} 1 & c \\ 1 & d \end{bmatrix}. \tag{9}$$

Then

$$C = \begin{bmatrix} KA & A \\ KF & -F \end{bmatrix} = \begin{bmatrix} KAF^{-1} & 0 \\ K & -I \end{bmatrix} \begin{bmatrix} F & FA^{-1}K^{-1}A \\ 0 & F + KFA^{-1}K^{-1}A \end{bmatrix}. \quad (10)$$

Since A, F, and K are nonsingular, C is singular only when $F + KFA^{-1}K^{-1}A = (FA^{-1}K + KFA^{-1})K^{-1}A$ is singular or when $G \stackrel{\text{def}}{=} FA^{-1}K + KFA^{-1}$ is singular. We determine that

$$G = \frac{1}{a-b} \begin{bmatrix} -2k'(b+c) & (1+k')(a+c) \\ -(1+k')(b+d) & 2(a+d) \end{bmatrix}. \quad (11)$$

Let $\tau = \frac{2(a+d)(b+c)}{(a+c)(b+d)}$. Then $\det(G) = 0$ when k' satisfies the quadratic equation:

$$k'^2 - 2(\tau - 1)k' + 1 = 0. \quad (12)$$

Now define the positive quantity

$$m \stackrel{\text{def}}{=} \frac{2(b-a)(d-c)}{(a+c)(b+d)}. \quad (13)$$

It is easily shown that $\tau - 1 = m + 1$ and the solution to Eq. 12 which is less than unity is

$$k' = \frac{1}{1 + m + \sqrt{m(2+m)}}. \quad (14)$$

The other solution is its reciprocal, which is greater than unity. From Eq. 10,

$$\begin{bmatrix} F & FA^{-1}K^{-1}A \\ \mathbf{0} & GK^{-1}A \end{bmatrix} \begin{bmatrix} \alpha \\ \gamma \\ \beta \\ \delta \end{bmatrix} = \mathbf{0} \text{ and } GK^{-1}A \begin{bmatrix} \beta \\ \delta \end{bmatrix} = \mathbf{0}. \quad (15)$$

We have

$$GK^{-1}A = \begin{bmatrix} (1+k')(a+c) - 2(b+c) & 2a(b+c) - b(1+k')(a+c) \\ -\frac{1+k'}{k'}(b+d) + 2(a+d) & \frac{1+k'}{k'}a(b+d) - 2b(a+d) \end{bmatrix}. \quad (16)$$

We now define $\sigma = 2(a+d)/(b+d)$ and obtain from the second row of Eq. 16:

$$[-(1+k') + \sigma k']\beta + [a(1+k') - b\sigma k']\delta = 0. \quad (17)$$

We preempt division by zero by setting

$$\delta = (1 + k' - \sigma k') \text{ and } \beta = a(1+k') - b\sigma k'. \quad (18)$$

The first row of C in Eq. 10 yields the relationship

$$KA\begin{bmatrix}\alpha\\ \gamma\end{bmatrix} + A\begin{bmatrix}\beta\\ \delta\end{bmatrix} = \mathbf{0}\,, \tag{19}$$

from which we obtain

$$k'(a\gamma - \alpha) = \beta - a\delta \text{ and } (b\gamma - \alpha) = \beta - b\delta\,. \tag{20}$$

Substituting the values for β and δ given in Eq. 18, we get

$$\alpha = b\sigma - a(1+k') \text{ and } \gamma = \sigma - (1+k')\,. \tag{21}$$

We must show that the transformation matrices B_1 and B_2 are nonsingular or that $\alpha\,\delta - \beta\,\gamma \neq 0$ for any intervals $[a,b]$ and $[c,d]$ for which $a+c>0$. We have

$$\begin{aligned}\alpha\,\delta - \beta\,\gamma &= [b\sigma - a(1+k')](1+k'-\sigma k') - [a(1+k') - b\sigma k'][\sigma - (1+k')]\\ &= \sigma(b-a)(1-k'^2) > 0\end{aligned} \tag{22}$$

We must also show that B_1 transforms the interior of $[k',1]$ into the interior of $[a,b]$ and that B_2 transforms the interior of $[k',1]$ into the interior of $[c,d]$. Since the transformations were generated to transform the endpoints properly, we need only show that one point x' outside of $[k',1]$ is such that $B_1(x')$ is outside $[a,b]$ and one point y' outside $[k',1]$ is such that $B_2(y')$ is outside $[c,d]$. First, we consider the case where $\gamma = 0$. We have $\sigma = 1+k', \alpha = (b-a)(1+k'), \delta = 1-k'^2$, and $\beta = (1+k')(a-bk')$. It follows that

$$B_1(x') = \frac{(b-a)(1+k')x' + (1+k')(a-bk')}{(1-k'^2)} = \frac{(b-a)x' + (a-bk')}{(1-k')}. \tag{23}$$

Thus, $x' = \infty$ transforms into $x = \infty$. The corresponding expression for $B_2(y')$ differs only in a negative sign for the second term in the numerator. Thus $y' = \infty$ transforms into $y = \infty$. The case of $\gamma = 0$ is thus resolved. When $\gamma \neq 0$, we choose $x' = -\frac{\delta}{\gamma}$ and $y' = \frac{\delta}{\gamma}$ so that $B_1(x') = \infty$ and $B_2(y') = \infty$. We then obtain from Eqs. 18 and 21:

$$\left|\frac{\delta}{\gamma}\right| = \left|\frac{1+k'-\sigma k'}{1+k'-\sigma}\right| = \left|\frac{1}{1-\frac{\sigma(1-k')}{1+k'-\sigma k'}}\right|. \tag{24}$$

This is greater than unity when $0 < \frac{\sigma(1-k')}{1+k'-\sigma k'} < 2$. We note from the definition of σ that $0<\sigma<2$. It follows that $1+k'-\sigma k' > 1-k' > 0$ and hence that

$$0 < \frac{\sigma(1-k')}{1+k'-\sigma k'} < \frac{\sigma(1-k')}{1-k'} = \sigma < 2, \tag{25}$$

as was to be shown. We have proved that the points at infinity for x and y correspond to points outside $[k', 1]$ and hence that $B_1([k', 1]) = [a, b]$ and $B_2([k', 1]) = [c, d]$.

The formulas derived here are of a simpler form than those given in [Wachspress, 1963, 1966]. The two formulations do however give identical iteration parameters. The iteration parameters for J iterations over the interval $[k', 1]$ are $w_j = dn[(2j - 1)K/2J, k]$. To determine p_j and q_j from w_j, we equate the roots of $g(x, y, \mathbf{p}, \mathbf{q})$ and $g(x', y', \mathbf{w}, \mathbf{w})$ to obtain $x' - w_j = B_1^{-1}(x) - w_j = 0$ when $x = B_1(w_j) = q_j$ and $y' - w_j = B_2^{-1}(y) - w_j = 0$ when $y = B_2(w_j) = p_j$. Thus,

$$p_j = \frac{\alpha w_j - \beta}{-\gamma w_j + \delta}, \text{ and } q_j = \frac{\alpha w_j + \beta}{\gamma w_j + \delta}. \tag{26}$$

The possibility of significant gain in efficiency is illustrated by the following example: Let the intervals be $[0.01, 10]$ and $[100, 1000]$. For Eq. 2 of Chap. 1, we would use $k' = \frac{0.01}{1000}$ which yields J varying as $\ln \frac{4}{k'} = 12.9$. The transformation equations yield

$$m = \frac{2(10 - 0.01)(1000 - 100)}{(0.01 + 100)(10 + 1000)} = 0.17802,$$

$$k' = \frac{1}{1 + m + \sqrt{m(2 + m)}} = 0.555.$$

Now $\ln \frac{4}{k'} = 1.97$ and the number of iterations is reduced by a factor of $\frac{12.9}{1.97} = 6.53$.

We also note that the generalized formulation only requires that matrix A be SPD. This ensures $a + c > 0$ and allows a splitting with either a or c less than zero. Convergence rate and relationships among J, k', and R are established in the transformed space.

2.3 The Three-Variable ADI Problem

Analysis of ADI iteration for three space variables is less definitive. Let X, Y, Z be the commuting components of the matrix A which are associated with line sweeps parallel to the x, y, z axes, respectively. Douglas (1962) proposed the iteration

$$(X + p_j I)\mathbf{u}_{j-2/3} = -2\left(Y + Z + \frac{X}{2} - \frac{p_j}{2} I\right)\mathbf{u}_{j-1} + 2\mathbf{b}, \tag{27.1}$$

$$(Y + p_j I)\mathbf{u}_{j-1/3} = Y\mathbf{u}_{j-1} + p_j \mathbf{u}_{j-2/3}, \tag{27.2}$$

$$(Z + p_j I)\mathbf{u}_j = Z\mathbf{u}_j + p_j \mathbf{u}_{j-1/3}. \tag{27.3}$$

Although Douglas suggested methods for choosing parameters 30 years ago, I am unaware at this time of any determination of optimum parameters as a function of spectral bounds. Moreover, error reduction as a function of parameter choice is not easily computed a priori. Perhaps a thorough literature search would uncover more extensive analysis. Rather than pursue this approach, we shall consider an alternative which allows a more definitive analysis.

Two of the three commuting matrices may be treated jointly. Let these be designated as H and V and let the third be P. We wish to solve the system

$$A\mathbf{u} \equiv (H + V + P)\mathbf{u} = \mathbf{b}. \tag{28}$$

The standard ADI iteration

$$(H + V + p_j I)\mathbf{u}_{j-1/2} = (p_j I - P)\mathbf{u}_{j-1} + \mathbf{b} \tag{29.1}$$

$$(P + q_j)\mathbf{u}_j = (q_j I - H - V)\mathbf{u}_{j-1/2} + \mathbf{b} \tag{29.2}$$

applies when solution of Eq. 29.1 is expedient, but this is not often the case. The analysis is simplified when applied in the transformed space where the eigenvalue intervals of $X' \equiv H' + V'$ and of $Z' \equiv P'$ are both $[k', 1]$. In this space the iteration parameters for the two sweeps are the same, and Eqs. 29 become

$$(X' + w_j I)\mathbf{u}_{j-1/2} = (w_j I - Z')\mathbf{u}_{j-1} + \mathbf{b}, \tag{30.1}$$

$$(Z' + w_j)\mathbf{u}_j = (w_j I - X')\mathbf{u}_{j-1/2} + \mathbf{b}. \tag{30.2}$$

Suppose we approximate $\mathbf{u}_{j-1/2}$ by standard ADI iteration applied to the commuting matrices $(H' + \frac{w_j}{2} I)$ and $(V' + \frac{w_j}{2} I)$. If this "inner" ADI iteration matrix is T_j, then Eq. 30.1 is replaced by

$$\mathbf{u}_{j-1/2} = T_j \mathbf{u}_{j-1} + (I - T_j)(X' + w_j I)^{-1}[(w_j I - Z')\mathbf{u}_{j-1} + \mathbf{b}]. \tag{31}$$

The error vector $\mathbf{e}_j \equiv \mathbf{u}_j - \mathbf{u}$ after the double sweep of Eqs. 30 is $L_j \mathbf{e}_{j-1}$, where

$$L_j = (Z' + w_j I)^{-1}(X' + w_j I)^{-1}(w_j I - X')[(w_j I - Z) + T_j(X' + Z')]. \tag{32}$$

T_j commutes with X' and Z'. Let the error reduction of the inner ADI iteration be ε_j. If this value is not sufficiently small, the iteration can diverge. This is illustrated by considering a limiting case of the eigenvector whose X'-eigenvalue is 1 and whose Z'-eigenvalue is k'. The corresponding eigenvalue of T_j is ε_j. The corresponding eigenvalue of L_j is

$$\lambda = \frac{(w_j - 1)[w_j - k' + \varepsilon_j(1 + k')]}{(w_j + 1)(w_j + k')}. \tag{33}$$

For one of the outer ADI iterations, w_j can be close to k' and thus small compared to unity. We consider the case where $w_j \doteq k'$. Then $|\lambda| \doteq \frac{\varepsilon_j}{2k'}$. We observe that ε_j must be less than $2k'$ for this eigenvalue to be less than unity. The composite J-step outer ADI iteration may still converge, but convergence can be seriously hampered by insufficient convergence of the inner ADI iteration. When sufficient inner ADI iterations are performed to ensure $\|T_j\| < 2k'$ for all j, the norm of the composite ADI iteration is bounded by the square root of the value achieved with Eq. 29. This is due to the factor of $(X' + w_j I)^{-1}(w_j I - X')$ in Eq. 32. In Chap. 3 we shall discuss use of ADI iteration as a preconditioner for a conjugate gradient iteration. In this application, modest error reduction is required of the ADI iteration.

The three-variable ADI iteration is not performed in the transformed space, and the analysis leading to Eqs. 32–33 must be modified accordingly. We find that with $X = H + V$ and $Z = P$ Eq. 32 becomes

$$L_j = (Z + q_j I)^{-1}(X + p_j I)^{-1}(q_j I - X)[(p_j I - Z) + T_j(X + Z)]. \quad (32A)$$

Applying the WBJ transformation to this equation, we find that Eq. 33 becomes

$$\lambda = \frac{(w_j - x)}{(w_j + x)}\left[(1 - \varepsilon_j)(w_j - z) + \varepsilon_j(w_j + x)\frac{(\delta - \gamma z)}{(\delta + \gamma x)}\right]. \quad (33A)$$

A careful analysis of the spectrum reveals that the square root of the convergence rate attained by Eq. 29 is guaranteed when

$$\varepsilon_j < \min\left[\frac{w_j(\delta + \gamma)}{(\delta - \gamma w_j)}, \frac{2k'(\delta + \gamma)}{(1 + k')(\delta - \gamma w_j)}\right]. \quad (34)$$

This bound on ε_j is approximately equal to the smaller of $2k'$ and w_j. This iteration does not appear to be particularly efficient when significant error reduction is required as a result of the many H, V iterations for each P-step. We defer further analysis until after we have discussed ADI preconditioning for conjugate gradients in Chap. 3.

2.4 Analysis of the Two-Variable Minimax Problem[2]

We consider the spectral radius of the generalized ADI equations (Eq. 3 of Chap. 1) after Jordan's transformation. Let $a_j \equiv p'_j$ and $b_j \equiv q'_j$. Then

$$\rho(G_J) = \max_{k' \le x, y \le 1}\left|\prod_{j=1}^{J}\frac{(b_j - x)(a_j - y)}{(a_j + x)(b_j + y)}\right|.$$

[2]Shortly after my book on "Iterative Solution of Elliptic Systems" was published, I received a phone call from Bruce Kellogg (University of Maryland) asking if anyone had ever solved the two-variable ADI minimax problem. I thought that Bill Jordan and I had done so. After all, it was obvious from symmetry considerations, after Jordan's transformation to yield identical ranges for the two variables, that the two-variable solution to the transformed problem was equal to the one-variable solution. Or was it obvious? After careful consideration, I determined that it was not evident and that, in fact, I could find no simple proof. I spent a good deal of time on this problem during the summer of 1967 and the analysis was of sufficient depth that I submitted it as my RPI PhD thesis, from which this section has been extracted. The thesis flavor is retained by the attention to detail here.

We consider the three parts of Chebyshev minimax analysis: existence, alternance, and uniqueness. We first note that if any a_j or b_j is less than k', then replacing that value by k' will decrease the magnitude of each nonzero factor in this product. Similarly, replacing any a_j or b_j greater than unity by unity will also decrease the magnitude of each nonzero factor. In our search for optimum parameters, we may restrict them to lie in the interval $[k', 1]$. When all the parameters are in this interval each factor has magnitude less than unity, and hence $\rho < 1$. Once it is shown that ρ is a continuous function of the parameters, standard compactness arguments may be used to establish the existence of a solution to the minimax problem.

The spectral radius is not affected by any change in the order in which the parameters are applied. We choose the nondecreasing ordering: $a_j \leq a_{j+1}$ and $b_j \leq b_{j+1}$. It will be demonstrated eventually that the optimum parameters in each set are distinct. Uniqueness will then be established for the ordered optimum parameter sets. In the ensuing analysis all parameter sets are restricted to the interval $[k', 1]$. We now establish continuity of ρ. We define $g(x, \mathbf{a}, \mathbf{b})$ as

$$g(x, \mathbf{a}, \mathbf{b}) = \prod_{j=1}^{J} \frac{a_j - x}{b_j + x}. \tag{35}$$

Then

$$\rho(G_J) = \max_{k' \leq x, y \leq 1} |g(x, \mathbf{a}, \mathbf{b}) g(y, \mathbf{b}, \mathbf{a})|. \tag{36}$$

Let $Z = \max_j |z_j|$ for any J-tuple $\mathbf{z}$. Consider a perturbation from parameter sets $\mathbf{a}$ and $\mathbf{b}$ to $\mathbf{a} + \mathbf{c}$ and $\mathbf{b} + \mathbf{f}$. Let $\rho(\mathbf{a}, \mathbf{b})$ be attained at (x_1, y_1) and let $\rho(\mathbf{a} + \mathbf{c}, \mathbf{b})$ be attained at (x_2, y_2), where $\rho(\mathbf{a}, \mathbf{b}) \leq \rho(\mathbf{a} + \mathbf{c}, \mathbf{b})$. (The argument is similar if the reverse inequality is assumed.) Since g is uniformly continuous over $[k', 1]^{2J+1}$, there exists for any $e > 0$ a $d > 0$ such that $|g(x_2, \mathbf{a} + \mathbf{c}, \mathbf{b}) g(y_2, \mathbf{b}, \mathbf{a} + \mathbf{c}) - g(x_2, \mathbf{a}, \mathbf{b}) g(y_2, \mathbf{b}, \mathbf{a})| < e/2$ for any $\mathbf{c}$ for which $C < d$.

For any real numbers w and u, $\big||w| - |u|\big| \leq |w - u|$. Thus,

$$\big|\rho(\mathbf{a} + \mathbf{c}, \mathbf{b}) - |g(x_2, \mathbf{a}, \mathbf{b}) g(y_2, \mathbf{b}, \mathbf{a})|\big| < e/2.$$

Moreover,

$$\rho(\mathbf{a} + \mathbf{c}, \mathbf{b}) \geq \rho(\mathbf{a}, \mathbf{b}) \geq |g(x_2, \mathbf{a}, \mathbf{b}) g(y_2, \mathbf{b}, \mathbf{a})|.$$

Therefore, when $C < d$,

$$|\rho(\mathbf{a} + \mathbf{c}, \mathbf{b}) - \rho(\mathbf{a}, \mathbf{b})| \leq \big|\rho(\mathbf{a} + \mathbf{c}, \mathbf{b}) - |g(x_2, \mathbf{a}, \mathbf{b}) g(y_2, \mathbf{b}, \mathbf{a})|\big| < e/2.$$

Similarly, there is an $h > 0$ such that when $F < h$, we have

$$|\rho(\mathbf{a} + \mathbf{c}, \mathbf{b} + \mathbf{f}) - \rho(\mathbf{a} + \mathbf{c}, \mathbf{b})| < e/2.$$

Therefore, when $C < d$ and $F < h$,

$$\begin{aligned}|\rho(\mathbf{a}+\mathbf{c},\mathbf{b}+\mathbf{f})-\rho(\mathbf{a},\mathbf{b})| &= |\rho(\mathbf{a}+\mathbf{c},\mathbf{b}+\mathbf{f})-\rho(\mathbf{a}+\mathbf{c},\mathbf{b})+\rho(\mathbf{a}+\mathbf{c},\mathbf{b})-\rho(\mathbf{a},\mathbf{b})| \\ &\leq |\rho(\mathbf{a}+\mathbf{c},\mathbf{b}+\mathbf{f})-\rho(\mathbf{a}+\mathbf{c},\mathbf{b})|+|\rho(\mathbf{a}+\mathbf{c},\mathbf{b})-\rho(\mathbf{a},\mathbf{b})| \\ &< e.\end{aligned}$$

Thus, $\rho(\mathbf{a},\mathbf{b})$ is continuous over $[k',1]^{2J}$ and it follows that ρ must attain its minimum value over $[k',1]^{2J}$ for at least one pair of J-tuples. We have established the existence of a solution to the two-variable ADI minimax problem, and we now address the alternance property. In the ensuing discussion, $\mathbf{a}^o$ and $\mathbf{b}^o$ are J-tuples for which ρ attains its least value and perturbations in the analysis are restricted so that all components remain in $[k',1]$. We will prove the following theorem:

Theorem 5 (The two-variable Poussin Alternance Property). *If*

$$\rho(\mathbf{a}^o,\mathbf{b}^o) = \min_{\mathbf{a},\mathbf{b}} \rho(\mathbf{a},\mathbf{b}),$$

then both $g(x,\mathbf{a}^o,\mathbf{b}^o)$ and $g(y,\mathbf{b}^o,\mathbf{a}^o)$ attain their maximum absolute values with alternating signs $J+1$ times on $[k',1]$.

The proof is long, and we require three lemmas:

Lemma 6. *The components of $\mathbf{a}^o$ are distinct and the components of $\mathbf{b}^o$ are distinct.*

Proof. We show that the assumption $a_k^o = a_{k+1}^o$ leads to a contradiction. The identical argument applies to $\mathbf{b}^o$. Let

$$G = \max_{k'\leq x\leq 1} |g(x,\mathbf{a}^o,\mathbf{b}^o)| \tag{37}$$

and

$$H = \max_{k'\leq y\leq 1} |g(y,\mathbf{b}^o,\mathbf{a}^o)|. \tag{38}$$

Then $\rho(\mathbf{a}^o,\mathbf{b}^o) = GH$. Let $P(x) = 1$ when $J = 2$ and for $J > 2$ define the polynomial

$$P(x) = \prod_{\substack{j=1 \\ j\neq k,k+1}}^{J} (a_j^o + x).$$

Now consider

$$g(x,\mathbf{a}_e,\mathbf{b}^o) = \frac{\prod_{j=1}^J (a_j^o - x) - exP(-x)}{\prod_{j=1}^J (b_j^o + x)}, \tag{39}$$

where e is a positive number which will subsequently be defined more precisely and where a_e is the J-tuple whose components are the zeros of the numerator on the right-hand side. The value of e is chosen sufficiently small that all these zeros are positive. These zeros include the $J-2$ roots of $P(-x)$ and the two roots in $[k', 1]$ of the quadratic $(a_k^o - x)^2 - ex = 0$. For all components of $\mathbf{a}^o$ in $[k', 1]$ and e positive, this quadratic has two real positive roots. Hence, all J roots are positive.

In general, $g(x, \mathbf{b}, \mathbf{a}) = g(-x, \mathbf{a}, \mathbf{b})^{-1}$. Hence,

$$g(y, \mathbf{b}^o, \mathbf{a}_e) = \frac{\prod_{j=1}^{J}(\mathbf{b}_j^o - y)}{\prod_{j=1}^{J}(\mathbf{a}_j^o + y) + eyP(y)}, \tag{40}$$

where both terms in the denominator are positive when e, y and all components of $\mathbf{a}^o$ are positive. Therefore, if we define

$$H_e \equiv \max_{k' \le y \le 1} |g(y, \mathbf{b}^o, \mathbf{a}_e)|, \tag{41}$$

then $H_e < H$. We next define

$$z(x) = g(x, \mathbf{a}^o, \mathbf{b}^o) - g(x, \mathbf{a}_1, \mathbf{b}^o) = \frac{xP(-x)}{\prod_{j=1}^{J}(\mathbf{b}_j^o + x)}. \tag{42}$$

(We note that when $e = 1$, $\mathbf{a}_e = \mathbf{a}_1$.) We observe that $g(x, \mathbf{a}_e, \mathbf{b}^o) = g(x, \mathbf{a}^o, \mathbf{b}^o) - ez(x)$. When all components of $\mathbf{a}^o$ and x are in $[k', 1]$, $|a_j^o - x| < 1$ and $|b_j^o + x| \ge 2k'$. Thus, if we define $M \equiv 2(2k')^{-J}$ then $|z(x)| < M$. Let $e_o = G/M$. Then, $0 < e|z(x)| < G$ when $z(x) \ne 0$ and $g(x, \mathbf{a}^o, \mathbf{b}^o) = 0$ when $z(x) = 0$. Moreover, $\operatorname{sign} g(x, \mathbf{a}^o, \mathbf{b}^o) = \operatorname{sign} z(x)$ when $g \ne 0$. It follows that

$$\begin{aligned} \left|g(x, \mathbf{a}_e, \mathbf{b}^o)\right| &= \left|g(x, \mathbf{a}^o, \mathbf{b}^o) - ez(x)\right| \\ &= \left||g(x, \mathbf{a}^o, \mathbf{b}^o)| - e|z(x)|\right| < G. \end{aligned} \tag{43}$$

If we define $G_e \equiv \max_{k' \le x \le 1} |g(x, \mathbf{a}_e, \mathbf{b}_o)|$, then $G_e < G$. We have already shown that $H_e < H$. Hence, $G_e H_e < GH = \rho(\mathbf{a}^o, \mathbf{b}^o)$, in contradiction to the hypothesis that the latter is a lower bound on the spectral radius. This establishes the lemma.

We next prove

Lemma 7. *If G and H are as defined in Lemma 6:*

i. $g(k', \mathbf{a}^o, \mathbf{b}^o) = G$ *and* $g(k', \mathbf{b}^o, \mathbf{a}^o) = H$
ii. $g(1, \mathbf{a}^o, \mathbf{b}^o) = (-1)^J G$ *and* $g(1, \mathbf{b}^o, \mathbf{a}^o) = (-1)^J H$

Proof. The components of the J-tuples $\mathbf{a}^o$ and $\mathbf{b}^o$ are in $[k', 1]$ so that if we define V by $g(k', \mathbf{a}^o, \mathbf{b}^o) = G - V$, then $0 \le V \le G$. Let $\mathbf{a}'$ differ from $\mathbf{a}^o$ only in its first element: $a_1' = a_1^o + e$ with $e \in [0, e_o]$, where e_o is a nonnegative number to be defined. Let $G' \equiv \max_{k' \le x \le 1} |g(x, \mathbf{a}', \mathbf{b}^o)|$, and let $H' \equiv \max_{k' \le x \le 1} |g(y, \mathbf{b}^o, \mathbf{a}')|$. Let $e_1 \equiv a_2^o - a_1^o$. By Lemma 6, $e_1 > 0$. Excluding the values $x = a_j^o$ for $j = 2, 3, \ldots, J$

and $y = b_j^o$ for $j = 1, 2, \ldots, J$, where $g(x, \mathbf{a}', \mathbf{b}^o) = g(y, \mathbf{b}^o, \mathbf{a}') = 0$, we have for $x \geq a_1^o + e$ and $y \in [k', 1]$,

$$\left|\frac{g(x, \mathbf{a}', \mathbf{b}^o)g(y, \mathbf{b}^o, \mathbf{a}')}{g(x, \mathbf{a}^o, \mathbf{b}^o)g(y, \mathbf{b}^o, \mathbf{a}^o)}\right| = \left|\frac{(x - a_1^o - e)(y + a_1^o)}{(x - a_1^o)(y + a_1^o + e)}\right| < 1. \tag{44}$$

Therefore,

$$\max_{a_1^o + e \leq x \leq 1, k' \leq y \leq 1} |g(x, \mathbf{a}', \mathbf{b}^o)g(y, \mathbf{b}^o, \mathbf{a}')| < \max_{k' \leq x, y \leq 1} |g(x, \mathbf{a}^o, \mathbf{b}^o)g(y, \mathbf{b}^o, \mathbf{a}^o)| = GH. \tag{45}$$

When $y = b_j^o$, $g(y, \mathbf{b}^o, \mathbf{a}') = g(y, \mathbf{b}^o, \mathbf{a}^o) = 0$ for $j = 1, 2, \ldots, J$. For all other $y \in [k', 1]$,

$$\left|\frac{g(y, \mathbf{b}^o, \mathbf{a}')}{g(y, \mathbf{b}^o, \mathbf{a}^o)}\right| = \left|\frac{y + a_1^o}{y + a_1^o + e}\right| < 1. \tag{46}$$

Hence,

$$H' < H. \tag{47}$$

For $k' \leq x \leq a_1$, $\frac{\partial\left|\frac{a_j - x}{b_j + x}\right|}{\partial x} = -\frac{a_j + b_j}{(b_j + x)^2} < 0$. Hence, $g(x, \mathbf{a}, \mathbf{b})$ increases in absolute value as x decreases from a_1 to k'. It follows that for $e \in (0, e_1)$,

$$\max_{k' \leq x \leq a_1^o + e} |g(x, \mathbf{a}', \mathbf{b}^o)| = g(k', \mathbf{a}', \mathbf{b}^o). \tag{48}$$

If we define $S \equiv \frac{\prod_{j=2}^J (a_j^o - k')}{\prod_{j=1}^J (a_j^o + k')}$ we have $g(k', \mathbf{a}', \mathbf{b}^o) = G - V + eS$. Suppose $V \neq 0$ and let $e_o = \min(e_1, V/2S)$. Then for $0 < e < e_o$,

$$g(k', \mathbf{a}', \mathbf{b}^o) < G - V + e_o S \leq G - \frac{V}{2}. \tag{49}$$

Combining Eqs. 45–47, we have $G'H' < GH = \rho(\mathbf{a}^o, \mathbf{b}^o)$, contrary to the hypothesis that $\rho(\mathbf{a}^o, \mathbf{b}^o)$ is a lower bound on the spectral radius. The contradiction is resolved only if $V = 0$, in which case $e_o = 0$ and $g(k', \mathbf{a}^o, \mathbf{b}^o) = G$. The same argument applied to $g(k', \mathbf{b}^o, \mathbf{a}^o)$ establishes that this is equal to H, and part (i) of the lemma is proved.

Part (ii) of the lemma can be proved by symmetry properties. Let $x = k'/x'$ and $y = k'/y'$. Then the minimax problem in terms of the primed variables is the same as the original problem with J-tuples related by: Components of $\mathbf{a}'$ equal components of $k'/\mathbf{a}$ in reverse order, and components of $\mathbf{b}'$ equal components of $k'/\mathbf{b}$ in reverse order. Since $g(x', \mathbf{a}'^o, \mathbf{b}'^o) = (-1)^J g(x, \mathbf{a}^o, \mathbf{b}^o)$ and $g(y', \mathbf{b}'^o, \mathbf{a}'^o) = (-1)^J g(y, \mathbf{b}^o, \mathbf{a}^o)$, part (ii) of the lemma is established by substituting k' for x in these equations. One reasons that if (ii) were not true for some minimizing set of parameters, then (i) would not be true in the primed system. But we have already established (i) for any minimizing set.

For a fixed pair of positive J-tuples, g is a rational function of x and is continuous for positive x. One more lemma will be proved before we establish the Chebyshev alternance property of the optimizing parameters. We first partition the interval $[k', 1]$ into subintervals such that $g(x)$ has only positive extrema, G, or only negative extrema, $-G$, with opposite signs in successive intervals. Since g can have at most J changes of sign, there can be at most $J + 1$ subintervals. Let g have only I alternations (i.e., $I + 1$ subintervals). Let the leftmost extreme point in subinterval $i + 1$ be $x_i(1)$ and the rightmost extreme point in this subinterval be $x_i(2)$. If there is only one extreme in the interval, $x_i(1) = x_i(2)$. By Lemma 7, $x_0(1) = k'$ and $x_I(2) = 1$. The function g is continuous over $[k', 1]$ and must therefore have at least one zero between $x_{i-1}(2)$ and $x_i(1)$. We choose any set of these zeros as u_i with $x_{i-1}(2) < u_i < x_i(1)$ for $i = 1, 2, \ldots, I$. There must be a positive V such that one of the following inequalities holds in each interval (u_i, u_{i+1}) for $i = 1, \ldots, I$:

$$-G + V < g(x) \leq G, \quad u_i \leq x \leq u_{i+1} \quad i \text{ even}, \tag{50.1}$$

$$-G \leq g(x) < G - V, \quad u_i \leq x \leq u_{i+1} \quad i \text{ odd}. \tag{50.2}$$

Similarly, if $h(y)$ has K alternations, we can select a set of v_k and a positive W such that for $k = 1, \ldots, K$:

$$-H + W < h(y) \leq H, \quad v_k \leq y \leq v_{k+1} \quad k \text{ even}, \tag{51.1}$$

$$-H \leq h(y) < H - W, \quad v_k \leq y \leq v_{k+1} \quad k \text{ odd}. \tag{51.2}$$

Let U be the smaller of V and W and define

$$F(x) \equiv -x \prod_{i=1}^{I} (u_i - x) \prod_{k=1}^{K} (v_k + x). \tag{52}$$

Since both $\mathbf{a}$ and $\mathbf{b}$ are positive, the products $\prod_{j=1}^{J}(a_j - x)$ and $\prod_{j=1}^{J}(b_j + x)$ have no common root. The Divisor Lemma in Chap. 1 establishes the existence of polynomials $P(x)$ and $R(x)$ of maximal degree J such that for $I + K + 1 \leq 2J$,

$$R(x) \prod_{j=1}^{J} (a_j - x) - P(-x) \prod_{j=1}^{J} (b_j + x) = F(x). \tag{53}$$

Since g and h can have at most J alternations in $[k', 1]$, $I + K + 1 > 2J$ if and only if $I = K = J$. It will be shown that this is indeed the case for any set of parameters for which ρ attains its lowest bound. If we assume to the contrary, we will find that polynomials P and R may be used to construct other sets of J-tuples for which the spectral radius is decreased. In the ensuing discussion, $\mathbf{a}$ and $\mathbf{b}$ are assumed to be optimal so that the conditions of Lemmas 6 and 7 are satisfied. Polynomials P and R satisfy Eq. 53 for these J-tuples. We are now ready to prove the final lemma:

Lemma 8. *Suppose g and h do not both have J Chebyshev alternations over $[k', 1]$. Then there is a positive value, e_0, such that for all $e \in (0, e_0)$ if we define*

$$g_1(x) = \frac{\prod_{j=1}^{J}(a_j - x) - eP(-x)}{\prod_{j=1}^{J}(b_j + x) - eR(x)} \tag{54.1}$$

and

$$h_1(y) = \frac{\prod_{j=1}^{J}(b_j - y) - eR(-y)}{\prod_{j=1}^{J}(a_j + y) - eP(y)}, \tag{54.2}$$

then

i. All the zeros of $g_1(x)$ and of $h_1(y)$ are real.
ii. $G_1 H_1 < GH$, where $G_1 = \max_{k' \le x \le 1} |g_1(x)|$ and $H_1 = \max_{k' \le y \le 1} |h_1(y)|$.

Proof. Let $N,\ X,\ Y,\ D$ be real numbers. When D and $D - Y$ are nonzero,

$$\begin{aligned}\frac{N - X}{D - Y} = \frac{D(N - X)}{D(D - Y)} &= \frac{D(N - X) + N(D - Y) - N(D - Y)}{D(D - Y)} \\ &= \frac{N}{D} + \frac{(NY - DX)}{D(D - Y)}.\end{aligned} \tag{55}$$

Applying this identity to g_1 and h_1, we get

$$g_1(x) = g(x) + \frac{eF(x)}{\prod_{j=1}^{J}(b_j + x)[\prod_{j=1}^{J}(b_j + x) - eR(x)]}, \tag{56.1}$$

and

$$h_1(y) = h(y) - \frac{eF(-y)}{\prod_{j=1}^{J}(a_j + y)[\prod_{j=1}^{J}(a_j + y) - eP(y)]}. \tag{56.2}$$

Let M be an upper bound on the magnitudes of the three polynomials $F(x),\ P(x)$, and $R(x)$ for $-1 \le x \le 1$. We note that $\prod_{j=1}^{J}(a_j + x)$ and $\prod_{j=1}^{J}(b_j + x)$ are each $\ge (2k')^J$. Let $e_1 = (2k')^J / M$. Then for $e \in (0, e_1)$ and $k' \le x, y \le 1$

$$\prod_{j=1}^{J}(b_j + x) - eR(x) \ge (2k')^J - eM > 0, \tag{57.1}$$

and

$$\prod_{j=1}^{J}(a_j + y) - eP(y) \ge (2k')^J - eM > 0. \tag{57.2}$$

From Eqs. 54–55, we conclude that

$$\mathrm{sign}[g_1(x) - g(x)] = \mathrm{sign} F(x), \tag{58.1}$$

$$\mathrm{sign}[h_1(y) - h(y)] = -\mathrm{sign} F(-y) \tag{58.2}$$

for $e \in (0, e_1)$ and $k' \leq x,\ y \leq 1$.

From the definition of $F(x)$ in Eq. 52, we obtain

$$F(x) < 0 \qquad u_i < x < u_{i+1} \text{ and } i \text{ even}\,, \tag{59.1}$$

$$F(x) > 0 \qquad u_i < x < u_{i+1} \text{ and } i \text{ odd}\,, \tag{59.2}$$

$$F(-y) > 0 \qquad v_k < y < v_{k+1} \text{ and } k \text{ even}\,, \tag{59.3}$$

$$F(-y) < 0 \qquad v_k < y < v_{k+1} \text{ and } k \text{ odd}\,. \tag{59.4}$$

Recalling the definition of U (after Eqs. 51) and of M (after Eqs. 56), we define

$$e_2' \equiv \frac{(2k')^{2J} U}{M[1 + (2k')^J U]} \text{ and } e_2 = \min(e_1, e_2'). \tag{60}$$

Then for $e \in (0, e_2)$ and $k' \leq x \leq 1$

$$|g_1(x) - g(x)| = \left| \frac{eF(x)}{\prod_{j=1}^{J}(b_j + x)[\prod_{j=1}^{J}(b_j + x) - eR(x)]} \right|$$

$$\leq \frac{eM}{(2k')^J[(2k')^J - eM]} < U \leq V. \tag{61}$$

Similarly, there is an e_3 such that for $e \in (0, e_3)$ and $k' \leq y \leq 1$, $|h_1(y) - h(y)| < U \leq W$. Let $e_4 = \min(e_2, e_3)$. For $e \in (0, e_4)$ and $k' = u_0 \leq x \leq u_1$, we have from Eq. 50

$$-G + V < g(x) \leq G. \tag{62.1}$$

By Eq. 59,

$$F(x) < 0, \tag{62.2}$$

and since sign $[g_1(x) - g(x)] = $ sign $F(x)$ is negative,

$$g_1(x) < g(x) \leq G. \tag{62.3}$$

Moreover, by Eq. 61, $|g_1(x) - g(x)| = g(x) - g_1(x) < U \leq V$ so that

$$g_1(x) > g(x) - V > -G. \tag{62.4}$$

From Eqs. 62.3 and 62.4, $-G < g_1(x) < G$. Also, $g(u_1) = F(u_1) = 0$. Hence, $g_1(u_1) = 0$. For $e \in (0, e_4)$ and $u_1 < x < u_2$, we have

$$-G \leq g(x) < G - V \text{ from Eq. 50,} \tag{63.1}$$

$$F(x) > 0 \text{ from Eq. 59,} \tag{63.2}$$

and sign $[g_1(x) - g(x)] =$ sign $F(x)$ is positive so that

$$g_1(x) > g(x) \geq -G. \tag{63.3}$$

Moreover, by Eq. 61, $|g_1(x) - g(x)| = g_1(x) - g(x) < U \leq V$. Hence,

$$g_1(x) < g(x) + V \leq G. \tag{63.4}$$

From Eqs. 63.3 and 63.4, $-G < g_1(x) < G$. Also, $g(u_2) = F(u_2) = 0$ so that $g_1(u_2) = 0$.

Continuing through all the intervals in this fashion, we find that $|g_1(x)| < G$ over $[k', 1]$. The same argument suffices to prove that $|h_1(y)| < H$. The lemma is thus proved.

The construction in proof of Lemma 8 fails only when $I = K = J$. Since $g_1(x)$ and $h_1(y)$ are continuous over $[k', 1]$, they can alternate J times over this interval only if all their zeros are in this interval. In fact, they are bounded rational functions in this interval whose numerators are polynomials of maximal degree J and accordingly have precisely J zeros in $[k', 1]$.

Since we have proved that a solution to the minimax problem exists, it follows immediately from Lemma 8 that for any J-tuples which achieve the least maximum there must be J Chebyshev alternations. We have thus proved:

Theorem 9 (Chebyshev alternance theorem). *Let* $\mathbf{a}^o$ *and* $\mathbf{b}^o$ *be* J*-tuples for which the spectral radius of the two-variable ADI error-reduction matrix is minimized. Then* $g(x, \mathbf{a}^o, \mathbf{b}^o)$ *and* $h(y, \mathbf{b}^o, \mathbf{a}^o)$ *both have* J *Chebyshev alternations on* $[k', 1]$.

Our final task is to establish uniqueness. Once we have proved that only one pair of ordered J-tuples can satisfy the Chebyshev theorem, we can assert that since the choice of $\mathbf{a} = \mathbf{b}$ equal to the optimizing J-tuple for the one-variable problem yields the Chebyshev alternance property, this choice is the unique solution to the two-variable problem.

Let $\mathbf{a}$ be the optimizing J-tuple for the one-variable problem with maximum value for $|g(x)|$ equal to G and let $\mathbf{a}', \mathbf{b}'$ be another set which yields the Chebyshev alternance property with maximum values for $|g'(x)|$ and $|h'(y)|$ equal to G' and H', respectively. We define the continuous function over $[k', 1]$:

$$d(x) \equiv g(x, \mathbf{a}, \mathbf{a}) - g(x, \mathbf{a}', \mathbf{b}') \equiv g(x) - g'(x). \tag{64}$$

When $G \neq G'$, it is easily shown that $d(x)$ alternates J times on $[k', 1]$ for if $G > G'$ then d has the sign of g at its alternation points and if $G < G'$ then d has the sign of g' at its alternation points. It follows that $d(x)$ has at least J zeros in $[k', 1]$.

When $G = G'$, the analysis is slightly more complicated. If $d(x) = 0$ at an interior alternation point, two sign changes are removed and only one zero identified at this alternation point. However, we note that the derivatives of both g and g' vanish at this common alternation point. Hence the derivative of d with respect to x also vanishes at this point and it is at least a double root. We thus recover the "lost" zero. Of course, the endpoint alternation points are common to both functions and each yields only one zero since the derivatives do not vanish at these points. However, each of these alternation points only accounts for one zero when $G \neq G'$. We have thus proved that $d(x)$ has at least J roots in $[k', 1]$ even when $G = G'$.

A similar argument applies to the difference between $h(y)$ and $h'(y)$. Now define

$$n(x) \equiv \prod_{j=1}^{J} (a_j - x)(b'_j + x) - \prod_{j=1}^{J} (a_j + x)(a'_j - x). \tag{65}$$

Then

$$d(x) = \frac{n(x)}{\prod_{j=1}^{J} (a_j + x)(b'_j + x)}. \tag{66}$$

Thus, since we have established that d has at least J zeros in $[k', 1]$, it follows that $n(x)$ has these same zeros. Applying the same argument to $h(y) - h'(y)$, we conclude that the polynomial

$$m(y) \equiv \prod_{j=1}^{J} (a_j - y)(a'_j + y) - \prod_{j=1}^{J} (a_j + y)(b'_j - y). \tag{67}$$

has at least J zeros in $[k', 1]$. We now observe that $n(-x) = -m(x)$. Therefore, the negatives of the zeros of $m(y)$ are also zeros of $n(x)$. Hence, $n(x)$ has at least $2J$ zeros. Inspection of Eq. 65 reveals that $n(x)$ is of maximal degree $2J - 1$. A contradiction is established unless $n(x)$ is the zero polynomial, in which case $\mathbf{a}' = \mathbf{b}' = \mathbf{a}$. We have proved the following:

Theorem 10 (Main Theorem). *The two-variable ADI minimax problem has as its unique solution the pair of J-tuples $\mathbf{a} = \mathbf{b}$ which are equal to the J-tuple that solves the one-variable ADI minimax problem.*[3]

[3] Having gone through this analysis, I was able to say in 1968 that it was indeed obvious that the optimum ADI parameters were the same for both sweeps in Eq. 3 of Chap. 1 when the spectral bounds for $F^{-1}H$ and $F^{-1}V$ were the same. Bill and I really did solve the two-variable ADI minimax problem back in 1963.

2.5 Generalized ADI Iteration[4]

The "GADI" iteration introduced in by Levenberg and Reichel in 1994 addresses possible improvement by performing a different number of sweeps in the two directions in each iteration. Their analysis is based on potential theory developed by Bagby (1969). There are two situations where GADI can outperform PR ADI (which they call CADI). One is where the work required to iterate in one direction is less than the work required in the other direction. They observe that this is the case for Sylvester's equation when the orders of matrices A and B (see Eqs. 3–66) differ significantly. Another example is the three-variable approach described in Sect. 2.3, where the H, V iteration even with one inner per outer requires twice the work of the P sweep. The second situation is where the two eigenvalue intervals differ appreciably. We will develop a more precise measure of this disparity.

Let

$$g(x,y) = \prod_{j=1}^{m} \frac{p_j - x}{p_j + y} \prod_{k=1}^{n} \frac{q_k - y}{q_k + x}. \tag{68}$$

We apply Jordan's transformation as described in Sect. 2.2 and find that

$$g(x,y,\mathbf{p},\mathbf{q}) = \left(\frac{\delta - \gamma y'}{\delta + \gamma x'}\right)^{m-n} g(x',y',\mathbf{p}',\mathbf{q}'). \tag{69}$$

When $m = n$ this reduces to the result of Sect. 2.2, but when $m \neq n$ there is an additional factor of

$$K_{m,n} = \left(\frac{\delta - \gamma y'}{\delta + \gamma x'}\right)^{m-n}. \tag{70}$$

When $\gamma = 0$, we have reduced the parameter optimization problem to one where both intervals are $[k', 1]$. We have already proved that in general $|\frac{\delta}{\gamma}| > 1$. If the work for the two directions is the same, we may choose $m \geq n$ when $\gamma > 0$ and $n \geq m$ when $\gamma < 0$. Then $K_{m,n}$ is in $(0, 1)$ for all x' and y' in $[k', 1]$. The following theorem establishes the preferential sweep direction in terms of the spectral intervals:

Theorem 11. *If the spectral interval for x is $[a,b]$ and for y is $[c,d]$, then $\gamma > 0$ if and only if $(d-c)(b+c) > (b-a)(a+d)$.*

---

[4]Periodically, my interest in ADI model-problem theory wanes. I see little need for further analysis. Then some new research area is uncovered and my enthusiasm is revived. One example is the discovery around 1982 of the applicability of ADI iteration to Lyapunov and Sylvester matrix equations. This led to need for generalization of the theory into the complex plane, a subject which will be covered in Chap. 4. In December of 1992 Dick Varga forwarded to me for comments and suggestions a draft of a paper by N. Levenberg and L. Reichel on "GADI" iteration. This "GADI" method differs from classical ADI (which they call CADI) in that one allows a different number of mesh sweeps in the two directions. This stimulated analysis presented here.

Proof. From the analysis in Sect. 2.2, we have

$$1 + k' = 2 + m - \sqrt{m(2+m)} = \tau - \sqrt{\tau(\tau-2)} = \tau\left[1 - \sqrt{1 - \frac{2}{\tau}}\right], \quad (71.1)$$

$$\frac{2}{\tau} = \frac{(a+c)(b+d)}{(a+d)(b+c)}, \quad (71.2)$$

$$\sigma = \frac{2(a+d)}{(b+d)} = \tau\frac{(a+c)}{(b+c)}, \quad (71.3)$$

$$\gamma = \sigma - (1+k') = \tau\left(\frac{(a+c)}{(b+c)} - 1 + \sqrt{1 - \frac{2}{\tau}}\right). \quad (71.4)$$

Since $\tau > 2$, we obtain from Eq. 71.4 $\gamma > 0$ when $(1 - \frac{2}{\tau}) > \left(\frac{b-a}{b+c}\right)^2$. Using Eq. 71.2 we find after a little algebra that this inequality reduces to $(d-c)(b+c) > (b-a)(a+d)$.

It follows that the greater number of sweeps should be in the direction of the variable with the larger normalized spectral interval. This is consistent with the potential analysis in Levenberg and Reichel.

In many applications $|\frac{\delta}{\gamma}| >> 1$ and $K_{m,n}$ are close to unity. It will now be shown that in this case CADI outperforms GADI when the work is the same in both directions. Let $G(m,n)$ be the maximum absolute value of $g(x', y', \mathbf{p}', \mathbf{q}')$ for the optimum parameter sets. Since x' and y' vary over the same interval, each value with $x' = y'$ occurs in g. The value for $G(n,m)$ must be greater than that attained with the optimum CADI parameters for $n + m$ sweeps. The CADI error reduction is $C(n+m) = G(n+m, n+m)^2$ for the corresponding $2(n+m)$ steps. Thus, $G(m,n) \geq \sqrt{C(n+m)}$. If the CADI asymptotic convergence rate is $\rho(C)$, then $C(s) \doteq \kappa\rho^s$ for some constant κ. The asymptotic convergence rate of GADI, $\rho(G)$, must therefore satisfy

$$\rho(G) = \lim_{m+n\to\infty} G(m,n)^{\frac{1}{m+n}} \geq \lim_{m+n\to\infty} C(n+m)^{\frac{1}{2(n+m)}} = \rho(C) \quad (72)$$

with equality only when $m = n$. One cannot anticipate significant improvement over CADI when the work is the same for the two ADI steps of each iteration and $K_{m,n} \doteq 1$. Any possible improvement arises from $K_{m,n}$ in Eq. 70, which can in certain circumstances render GADI more efficient. Suppose the y-direction is preferred ($\gamma > 0$). One strategy is to choose an integer value for r and let $m = rn$. Then the inequality in Eq. 72 becomes

$$\rho(G) \geq \frac{(\delta - \gamma)}{(\delta + \gamma)}\rho(C). \quad (73)$$

Even when K is close to unity, significant improvement may be achieved with GADI when the work differs for the two steps. As mentioned previously, this is true for the three-variable ADI iteration and for the Sylvester matrix equation when the orders of A and B differ appreciably. The minimax theory from which optimum CADI parameters were derived has not been generalized to GADI at this writing. The Bagby points described by Levenberg and Reichel do yield asymptotically optimal parameters. Their "generalized" Bagby points are easy to compute and provide a convenient means for choosing good parameters.

We leave GADI now and return to our discussion of "classical" ADI. The theory for determining optimum parameters and associated error reduction as a function of eigenvalue bounds for $F^{-1}H$ and $F^{-1}V$ is firm when these matrices commute and the sum of their lower bounds is positive. We first examine in Chap. 3 how to choose F to yield these "model problem" conditions for a class of elliptic boundary value problems. We then describe how this model problem may be used as a preconditioner for an even more general class of problems.

Chapter 3
Model Problems and Preconditioning

Abstract Model problem ADI iteration is discussed for three distinct classes of problems. The first is discretized elliptic systems with separable coefficients so that difference equations may be split into two commuting matrices. The second is where the model ADI problem approximates the actual nonseparable problem and serves as a preconditioner. The third is an entirely different class of problems than initially considered. These are Lyapunov and Sylvester matrix equations in which commuting operations are inherent.

3.1 Five-Point Laplacians

The heat-diffusion problem in two space dimensions was treated by Peaceman and Rachford (1955) in their seminal work on ADI iteration. They considered both time-dependent parabolic problems and steady-state elliptic problems. The Laplacian operator may be discretized over a rectangular region by standard differencing over a grid with spacing h in both the x and y directions. If one multiplies the equations by h^2, one obtains five-point interior equations with diagonal coefficient of 4 and off-diagonal coefficients of -1 connecting each interior node to its four nearest neighbors. Boundary conditions are incorporated in the difference equations. This is a model ADI problem when the boundary condition on each side is uniform. Given values need not be constant on a side, but one cannot have given value on part of the side and another condition like zero normal derivative on the remainder of the side. It was shown by Birkhoff, Varga and Young (1962) that there must be a full rectangular grid in order that model conditions prevail. For the Dirichlet problem (with values given on all boundaries), the horizontal coupling for a grid with m rows and n columns of unknowns when the equations are row-ordered is

$$H = \text{diag}_m[L_n], \tag{1.1}$$

$$L_n = \text{tridiag}_n[-1, 2, -1]. \tag{1.2}$$

E. Wachspress, *The ADI Model Problem*, DOI 10.1007/978-1-4614-5122-8_3,
© Springer Science+Business Media New York 2013

The subscripts designate the orders of the matrices. The vertical coupling is similar with m and n interchanged when the equations are column-ordered. When row-ordered this coupling is

$$V = \text{tridiag}_m[-I_n, 2I_n, -I_n], \tag{2}$$

where I_n is the identity matrix of order n. Matrices H and V commute and the simultaneous eigenvectors for $r = 1, 2, \ldots, m$ and $s = 1, 2, \ldots, n$ have components at the node in column i and row j of

$$v(r, s; i, j) = \sin \frac{ir\pi}{m+1} \sin \frac{js\pi}{n+1}. \tag{3}$$

The corresponding eigenvalues are

$$\lambda(H) = 2\left(1 - \cos \frac{s\pi}{n+1}\right), \tag{4.1}$$

$$\gamma(V) = 2\left(1 - \cos \frac{r\pi}{m+1}\right). \tag{4.2}$$

When the spacing is h along the x-axis and k along the y-axis, one may multiply the difference equations by the mesh-box area hk to yield matrices $H' = \frac{k}{h}H$ and $V' = \frac{h}{k}V$. The eigenvectors remain the same but the eigenvalues are now multiplied by these mesh ratios. It is seen that when the ratio of these increments (the "aspect ratio") differs greatly from unity, the spectra for the two directions differ significantly even when $m = n$. For optimal use of ADI iteration, one must consider the two-variable problem and apply Jordan's transformation to obtain parameters for use in the generalized equations, Eqs. 3 of Chap. 1.

Now consider variable increments, h_i between columns i and $i + 1$ and k_j between rows j and $j + 1$. The equation at node i, j may be normalized by the mesh-box area: $\frac{1}{4}(h_{i-1} + h_i)(k_{j-1} + k_j)$. Then

$$L_n = \text{tridiag}_n\left[-\frac{1}{h_{i-1}(h_{i-1} + h_i)}, \frac{1}{2h_{i-1}h_i}, -\frac{1}{h_i(h_{i-1} + h_i)}\right]. \tag{5}$$

Note that the elements of L_n do not depend on the row index j. The eigenvalues of matrix H are now the eigenvalues of tridiagonal matrix L_n, each of multiplicity m. The Jordan normal form of this matrix is diagonal since it is the product of a positive diagonal matrix and a symmetric matrix. Bounds on these eigenvalues must be computed in order to determine optimum iteration parameters. If the V matrix is ordered by columns, then the corresponding diagonal blocks of order m are tridiagonal matrices with k_j replacing h_i in Eq. 5. Thus, column-ordered $V = \text{tridiag}_n[S_m]$, with

$$S_m = \text{tridiag}_m\left[-\frac{1}{k_{j-1}(k_{j-1} + k_j)}, \frac{1}{2k_{j-1}k_j}, -\frac{1}{k_j(k_{j-1} + k_j)}\right]. \tag{6}$$

Eigenvalue bounds for S_m must also be estimated for determining iteration parameters. Instead of dividing the equations by the mesh-box areas, we may retain the H and V matrices so that $H + V$ is the difference approximation to the differential operator integrated over the mesh box. We now multiply the iteration parameters by the normalizing (diagonal) matrix F whose entries are the mesh-box areas. This approach has ramifications which are beneficial in a more general context. Iteration Eq. 4 of Chap. 1 yield a matrix whose eigenvectors are independent of the iteration parameters when $HF^{-1}V - VF^{-1}H = 0$. This is evidently true for this case where $F^{-1}H$ and $F^{-1}V$ commute. Commutation is revealed by the fact that the elements in $F^{-1}H$ (which are displayed in Eq. 5) depend only on the index i while the elements in $F^{-1}V$ (which are displayed in Eq. 6) depend only on the index j. The spectra for which parameters are computed remain those of $F^{-1}H$ and $F^{-1}V$.

The ADI model-problem conditions are attainable in any orthogonal coordinate system for a full rectangular grid. When the Laplacian operator is discretized by integrating over the mesh box around node ij, the diagonal matrix of mesh-box areas is the appropriate matrix F. In fact, the first application of ADI iteration with Eq. 3 of Chap. 1 included cylindrical and polar coordinates [Wachspress, 1957].

A comparison with Fast Fourier Transform solution of such problems is revealing [Concus and Golub, 1973]. When the spacing is uniform in each direction, the eigensolutions are known. When high accuracy is desired the FFT outperforms ADI in this case. However, when only modest error reduction is demanded ADI is quite competitive. The FFT suffers somewhat when the number of rows or columns is not a power of two, but that is more a programming complication than a deficiency of the approach. Now consider variable increments. For ADI iteration we need only eigenvalue bounds. For the FFT we need the complete eigensolutions for both the H and the V matrices. This is time-consuming, and ADI in general outperforms FFT in such cases. Only when the same grid is used with many forcing vectors can FFT become competitive in this more general case. There are other "Fast Poisson Solvers" which may outperform ADI when very high accuracy is demanded [Buzbee, Golub and Nielson, 1970].

Eigenvalue bounds for the tridiagonal matrices, L_n and S_m, are relatively easy to compute. The maximum absolute row sum provides an adequate upper bound. The iteration is insensitive to loose (but conservative) upper bounds. Lower bounds can be computed with shifted inverse iteration, starting with a guess of zero. There is only one tridiagonal matrix for each direction and the time for the eigenvalue bound computation is negligible compared to the iteration time.

3.2 The Neutron Group-Diffusion Equation

The neutron group-diffusion equation is

$$-\nabla \cdot D(x,y)\nabla u(x,y) + \sigma(x,y)u(x,y) = s(x,y), \tag{7}$$

where $D(x, y) > 0$ and $\sigma(x, y) \geq 0$. This is an ADI model problem when the region is rectangular with uniform boundary condition on each side and the coefficients are separable in that

$$D(x, y) = D(x)D'(y) \text{ and } \sigma(x, y) = D(x)D'(y)[\sigma(x) + \sigma'(y)], \tag{8}$$

for we may then divide the equation by $D(x)D'(y)$ and express the operator as the sum of two commuting operators, $\mathcal{H}$ and $\mathcal{V}$, where

$$\mathcal{H} = \frac{1}{D(x)} \frac{\partial}{\partial x} D(x) \frac{\partial}{\partial x} + \sigma(x) \tag{9.1}$$

and

$$\mathcal{V} = \frac{1}{D'(y)} \frac{\partial}{\partial y} D'(y) \frac{\partial}{\partial y} + \sigma'(y). \tag{9.2}$$

This is a slight generalization of the model problem displayed by Young and Wheeler (1964) in which σ was restricted to $KD(x)D'(y)$ with K constant.

When the neutron group-diffusion equation is discretized by the box-integration method, the difference forms of Eqs. 9 are each three-point equations. We need not divide the equations by $D(x, y)$ if we define the F matrix by

$$F = \text{diag}[(i, j)] = GG' = \text{diag}[g(i)]\,\text{diag}[g'(j)], \tag{10}$$

where

$$g(i) = \frac{1}{2}[D_i h_i + D_{i-1} h_{i-1}], \tag{11.1}$$

and

$$g'(j) = \frac{1}{2}[D'_j k_j + D'_{j-1} k_{j-1}]. \tag{11.2}$$

In these equations, $D_i = D(x)$ between columns i and $i + 1$ while $D'_j = D'(y)$ between rows j and $j + 1$. The coefficient matrix obtained by box-integration can now be expressed as

$$A = LG' + L'G, \tag{12}$$

where for row-ordered equations

$$L \equiv \text{diagonal}_m[L_n], \tag{13}$$

with the matrix L_n repeated as the m diagonal blocks in L given by

$$L_n = \text{tridiagonal}\left\{-\frac{D_{i-1}}{h_{i-1}}, \left[D_{i-1}\left(\frac{1}{h_{i-1}} + \frac{h_{i-1}\sigma_{i-1}}{2}\right) + D_i\left(\frac{1}{h_i} + \frac{h_i \sigma_i}{2}\right)\right], -\frac{D_i}{h_i}\right\}, \tag{14}$$

and for column-ordered equations

$$L' \equiv \text{diagonal}_n[L'_m], \tag{15}$$

with the matrix L'_m repeated as the n diagonal blocks in L' given by

$$L'_m = \text{tridiagonal}\left\{-\frac{D'_{j-1}}{k_{j-1}}, \left[D'_{j-1}\left(\frac{1}{k_{j-1}}+\frac{k_{j-1}\sigma'_{j-1}}{2}\right)+D'_j\left(\frac{1}{k_j}+\frac{k_j\sigma'_j}{2}\right)\right], -\frac{D'_j}{k_j}\right\}. \tag{16}$$

Here, σ_i is the value between columns i and $i+1$ while σ'_j is the value between rows j and $j+1$.

The primed and unprimed matrices of order mn commute. The ADI equations can be expressed in the form

$$(LG' + w_s GG')\mathbf{u}_{s-\frac{1}{2}} = -(L'G - w_s GG')\mathbf{u}_{s-1} + \mathbf{s}, \tag{17.1}$$

$$(L'G + w'_s GG')\mathbf{u}_s = -(LG' - w'_s GG')\mathbf{u}_{s-\frac{1}{2}} + \mathbf{s}, \tag{17.2}$$

$$s = 1, 2, \ldots, J.$$

The right-hand side of Eq. 17.1 may be computed with the column-ordered block diagonal matrix L' and column-ordered $\mathbf{u}$ and $\mathbf{s}$. The resulting vector may then be reordered by rows as the forcing term for Eq. 17.1 with row ordering. Similarly, the right-hand side of Eq. 17.2 may be computed in row order and transposed to column order.

Eigenvalue bounds must be computed for the commuting tridiagonal matrices $G_n^{-1}L_n$ and $G_m'^{-1}L'_m$ for determining optimum parameters and associated convergence. These matrices are similar to SPD matrices and methods described for the model Laplace equation suffice for computing these eigenvalue bounds.

3.3 Nine-Point (FEM) Equations

When the Laplace or neutron group-diffusion operator is discretized by the finite element method over a rectangular mesh with bilinear basis functions, the equations are nine-point rather than five-point. It is by no means obvious that these are model ADI problems. Although Peaceman and Rachford introduced ADI iteration in the 1950s and the theory relating to convergence and choice of optimum parameters was in place by 1963, it was not until 1983 that I discovered how to express the nine-point equations as a model ADI problem [Wachspress, 1984]. The catalyst for this generalization was the analysis of the generalized five-point model problem discussed in Sect. 3.2 and in particular the form of the ADI iteration in Eqs. 17. This method was first implemented in 1990 [Dodds, Sofu and Wachspress], roughly 45 years after the seminal work by Peaceman and Rachford. One might question the practical worth of such effort in view of the restrictions imposed by the model conditions. However, application of model-problem analysis to more general problems will be exposed in Sect. 3.4.

Finite element discretization is based on a variational principle applied with a set of basis functions over each element. The basis functions from which the nine-point equations over a rectangular grid are obtained are bilinear. These nine-point finite element equations are related to the five-point box-integration equations.

A detailed analysis reveals that when the model conditions of Eq. 8 are satisfied, the finite element equations can be expressed as in Eq. 12:

$$A\mathbf{u} \equiv (LG' + L'G)\mathbf{u} = \mathbf{s}, \tag{18}$$

where we define the unprimed matrices when the equations are ordered by rows as

$$L \equiv \text{diagonal}_m[L_n], \tag{19.1}$$

$$G \equiv \text{diagonal}_m[G_n], \tag{19.2}$$

with tridiagonal matrices repeated as diagonal blocks:

$$L_n = \text{tridiagonal}\left\{ D_{i-1}\left(\frac{h_{i-1}\sigma_{i-1}}{6} - \frac{1}{h_{i-1}}\right), \left[D_{i-1}\left(\frac{h_{i-1}\sigma_{i-1}}{3} + \frac{1}{h_{i-1}}\right) + D_i\left(\frac{h_i\sigma_i}{3} + \frac{1}{h_i}\right)\right], D_i\left(\frac{h_i\sigma_i}{6} - \frac{1}{h_i}\right)\right\} \tag{20}$$

and

$$G_n = \text{tridiagonal}[D_{i-1}h_{i-1},\ 2(D_{i-1}h_{i-1} + D_ih_i),\ D_ih_i]/6. \tag{21}$$

The primed matrices are of the same form when the equations are ordered by columns:

$$L' \equiv \text{diagonal}_n[L'_m], \tag{22.1}$$

$$G' \equiv \text{diagonal}_n[G'_m], \tag{22.2}$$

with tridiagonal matrices:

$$L'_m = \text{tridiagonal}\left\{ D'_{j-1}\left(\frac{k_{j-1}\sigma'_{j-1}}{6} - \frac{1}{k_{j-1}}\right), \left[D'_{j-1}\left(\frac{k_{j-1}\sigma'_{j-1}}{3} + \frac{1}{k_{j-1}}\right) + D'_j\left(\frac{k_j\sigma'_j}{3} + \frac{1}{k_j}\right)\right], D'_j\left(\frac{k_j\sigma'_j}{6} - \frac{1}{k_j}\right)\right\} \tag{23}$$

and

$$G'_m = \text{tridiagonal}[D'_{j-1}k_{j-1},\ 2(D'_{j-1}k_{j-1} + D'_jk_j),\ D'_jk_j]/6. \tag{24}$$

The σ terms in the L and L' matrices are characteristic of finite element rather than box-integration equations, but this difference is sometimes eliminated by the "lumped mass" finite element approach which reduces the σ contribution to the box-integration diagonal contribution. Matrices L_n and L'_m in Eqs. 20 and 23 are then identical to matrices L_n and L'_m in Eqs. 14 and 16. This has no effect on the ADI analysis. The G and G' matrices are now tridiagonal diffusion-coefficient-weighted Simpson rule quadrature matrices. The fact that these matrices are tridiagonal rather than diagonal seems to preclude efficient ADI iteration, but we shall soon show how this is remedied.

We consider the ADI-type iteration defined in Eq. 17:

$$(LG' + w_s GG')\mathbf{u}_{s-\frac{1}{2}} = -(L'G - w_s GG')\mathbf{u}_{s-1} + \mathbf{s}, \tag{25.1}$$

$$(L'G + w'_s GG')\mathbf{u}_s = -(LF' - w'_s GG')\mathbf{u}_{s-\frac{1}{2}} + \mathbf{s}, \tag{25.2}$$

$$s = 1, 2, \ldots, J.$$

Since G and G' are tridiagonal rather than diagonal, the systems to be solved in each step are not block tridiagonal but have the same structure as the coefficient matrix A. They are systems of nine-point equations. We must somehow reduce these iteration equations to the form of Eqs. 1–3 with tridiagonal systems on the left-hand sides. For this purpose we define the vectors

$$\mathbf{v}_{s-\frac{1}{2}} = G'\mathbf{u}_{s-\frac{1}{2}} \tag{26.1}$$

and

$$\mathbf{v}_s = G\mathbf{u}_s. \tag{26.2}$$

One starts the iteration by computing $\mathbf{v}_0 = G\mathbf{u}_0$ and by virtue of commutativity of primed and unprimed matrices rewrites Eqs. 25 as

$$(L + w_s G)\mathbf{v}_{s-\frac{1}{2}} = -(L' - w_s G')\mathbf{v}_{s-1} + \mathbf{s}, \tag{27.1}$$

$$(L' + w'_s G')\mathbf{v}_s = -(L - w'_s G)\mathbf{v}_{s-\frac{1}{2}} + \mathbf{s}, \tag{27.2}$$

$$s = 1, 2, \ldots, J.$$

These equations are almost the same as the five-point iteration equations. They differ only in that the iteration parameters are multiplied by tridiagonal rather than diagonal matrices. However, the matrices on each side of these equations have the same structure as the corresponding five-point matrices. The coefficient matrix on the left side of Eq. 27.1 for update of all rows is the tridiagonal matrix $(L_n + w_s G_n)$, and the coefficient matrix on the left side of Eq. 27.2 for update of all columns is the tridiagonal matrix $(L'_m + w'_s G'_m)$. The iteration is terminated with recovery of $\mathbf{u}_J$ after J iterations by solving the tridiagonal systems $G\mathbf{u}_J = \mathbf{v}_J$.

The eigenvalue bounds for $G_n^{-1} L_n$ and $G'^{-1}_m L'_m$ must be computed. These may be treated as generalized eigenvalue problems: $L_n\mathbf{e} = \lambda G_n \mathbf{e}$ and $L'_m \mathbf{e}' = \gamma G'_m \mathbf{e}'$. Shifted inverse iteration has been used to compute upper and lower bounds for these eigenvalues. Some simple observations facilitate the computation. Matrices L_n and L'_m have positive inverses [Varga, 1962] and matrices G_n and G'_m are irreducible and nonnegative. Therefore, matrices $L_n^{-1}G_n$ and $L'^{-1}_m G'_m$ are positive. The Perron theorem asserts that the largest eigenvalues of these matrices have positive eigenvectors. If we choose $\mathbf{e}_0$ as a vector with all components equal to unity and solve the tridiagonal systems $L_n\mathbf{e}_1 = G_n\mathbf{e}_0$ and $L'_m\mathbf{e}'_1 = G'_m\mathbf{e}_0$, then the largest components of $\mathbf{e}_1$ and $\mathbf{e}'_1$ are upper bounds on the largest eigenvalues of these positive matrices. Their reciprocals are therefore lower bounds for the smallest eigenvalues of $G_n^{-1}L_n$ and $G'^{-1}_m L'_m$, respectively. These bounds may be used as a first shift in the computation of the lower eigenvalue bounds. First estimates for upper bounds may be computed with Rayleigh quotients $\frac{\mathbf{f}_0^T, L_n\,\mathbf{f}_0}{\mathbf{f}_0^T, G_n\,\mathbf{f}_0}$ and $\frac{\mathbf{f}_0^T, L'_m\,\mathbf{f}_0}{\mathbf{f}_0^T, G'_m\,\mathbf{f}_0}$, where the components of $\mathbf{f}_0$ alternate between plus one and minus one.

3.4 ADI Model-Problem Limitations

We have described a class of boundary value problems to which ADI model-problem theory applies. There is no other iterative method for which precise convergence prediction is possible that has the logarithmic dependence on problem condition. (We measure problem condition of an SPD system by the ratio of maximum to minimum eigenvalue of the coefficient matrix. This condition often varies as the number of nodes in the grid when spacing retains the same uniformity as the grid is refined.) Preconditioned conjugate gradient and multigrid computation may be competitive and even superior for some of these problems, but convergence theory is less definitive. Successive overrelaxation and Chebyshev extrapolation converge as the square root of the condition of the problem. For moderately difficult ADI model problems, the ADI iteration is more efficient. For example, the five-point Laplace problem with equal spacing and a 100×100 grid requires about 150 SOR iterations and only 10 ADI iterations for an error reduction by a factor of 10^{-4}. One model-problem ADI iteration, including both sweeps, requires about twice the work of one SOR iteration, but ADI has a clear advantage here. This advantage tends to manifest itself with smaller grids when mesh spacing is not uniform.

The greatest failing of ADI iteration is not in solution of model problems, but rather in restrictions imposed by the model conditions. Practitioners often demand methods which are applicable to a greater variety of problems. ADI iteration is often applied to problems for which model conditions are not met. Although considerable success has been realized for a variety of problems, departure from model conditions can lead to significant deterioration of the rapid convergence characteristic of ADI applied to model problems. Varga (1962) illustrated this with a simple problem contrived so that ADI iteration diverges with parameters chosen as though model conditions are satisfied when in reality they are not. Theory relating to parameter selection for general problems is sketchy. Although convergence can be guaranteed with some choices, the rate of convergence can rarely be predicted with variable parameters when the model conditions are not satisfied. It is this lack of sound theoretical foundations that motivated restriction of this monograph to application of ADI iteration only to model problems. In the next section we describe how model-problem ADI iteration may be applied to solve problems for which model conditions are not satisfied.

3.5 Model-Problem Preconditioners

3.5.1 Preconditioned Iteration

Several significant concepts were Introduced in Wachspress (1963). The Peaceman–Rachford ADI equations (Eq. 1 of Chap. 1) were generalized with different parameters for the two sweeps each iteration (Eqs. 1–3) to improve efficiency in solution of problems with different spectral intervals for the two directions. The earlier AGM algorithm for computing parameters when $J = 2^n$ (Sect. 1.4) was extended to this generalized iteration. This algorithm motivated Jordan's transformation of variables (Sect. 2–1.3). Both the variable transformation and Jordan's elliptic-function solution to the minimax problem were published for the first time as an appendix in Wachspress (1963).

The method now known as "preconditioned conjugate gradients" was also introduced in this paper as "compound iteration." Studies performed in 1962 established the potency of this new procedure, but the sparse numerical studies reported in this paper stimulated little interest and the method lay dormant for several years. It was rediscovered, was enhanced with a variety of preconditioners, and is now one of the more universally used methods for solving large elliptic type systems.

Compound iteration with ADI inner iteration was introduced by D'Yakonov (1961) to extend application of model problem ADI iteration to problems for which the model conditions were violated. The model problem was thus used as a "preconditioner" for the true problem. The term preconditioner was not introduced until several years after D'Yakonov's paper appeared. D'Yakonov used a two-term "outer" iteration with a constant extrapolation that converged about the same as Gauss–Seidel applied to the preconditioned system. The combination of ADI preconditioning and Lanczos-type[1] outer iteration was the new aspect of the analysis in my 1963 paper. This is in general much more efficient than Gauss–Seidel iteration.

[1]Nowadays, a variety of names are attached to variants of the Lanczos recursion formulas derived by minimizing different functionals. Forty years ago Gabe Horvay (a GE mechanics expert and one of my associates at KAPL) introduced me to this new approach developed by his friend Lanczos and, influenced strongly by Gabe, I became accustomed to referring to all these schemes as "Lanczos algorithms." Hence, the method of "conjugate gradients" is often referred to as "Lanczos' method" in my early works.

The following description of compound iteration is taken directly from the 1963 paper. The wording parallels that of modern texts on this method. I have been unable to find an earlier published account of preconditioned conjugate gradients, and refer the reader to the comprehensive historical review by Golub and O'Leary (1987).

3.5.2 Compound Iteration (Quotations from Wachspress, 1963)

"Application of compound iteration with inner iteration other than ADI was described by Cesari (1937) and by Engeli et al. (1959). Use of ADI iteration in this manner was discussed first by D'Yakonov (1961). We wish to solve the matrix equation $A\mathbf{z} = \mathbf{s}$ for $\mathbf{z}$ when given the vector $\mathbf{s}$ and the real, positive definite matrix A.[2] It is not often possible to express A as the sum of two symmetric commuting matrices, H and V, such that the matrix inversions in [Eq. 3 of Chap. 1] are readily performed. There may, however, be a model problem matrix M which approximates A in the sense that $p(M^{-1}A) << p(A)$, where p is the p-condition number, equal in this case to the ratio of the maximum to minimum eigenvalues. The closer $p(M^{-1}A)$ is to unity, the more efficient compound iteration becomes."

The paper continued with proof of a theorem on the effect of termination of the ADI model problem iteration with error reduction ε on the condition of this compound iteration. The ADI iteration actually replaces M by an SPD matrix B_ε. A more detailed proof with useful innovations will be given in Sect. 3.6. The theorem asserts that the effective condition is

$$p(B_\varepsilon^{-1}A) \leq \frac{1+\varepsilon}{1-\varepsilon} p(M^{-1}A). \tag{28}$$

Next, details were given for a symmetric conjugate gradient algorithm applied directly to the system $B^{-1}A$. This was the first published account of applicability of this algorithm to a product of SPD matrices. Hestenes and Stieffel (1952) discussed preconditioning of nonsymmetric systems with their transposes to yield symmetric systems. The observation that these algorithms could be applied to a product of SPD matrices is trivial and can be cast as application to A with inner products defined as $(\mathbf{w}\,,\,\mathbf{z}) = \mathbf{w}^T B^{-1}\mathbf{z}$. When B is SPD one can define a norm consistent with this inner product as $\|\mathbf{u}\| = (\mathbf{u}\,,\,\mathbf{u})^{\frac{1}{2}}$. The conjugate gradient algorithm then minimizes the norm of the residual vector.

After giving the conjugate gradient algorithm inner products and recursion formulas, my 1963 paper continued with: "The number of Lanczos iterations for a prescribed error reduction varies as $\sqrt{p(B^{-1}A)}$. [A footnote attributed this result to Lanczos being at least as efficient as Chebyshev extrapolation.] To gain some

[2]My definition of positive definite in those days implied symmetry. More recently, the term has been used by some with a different definition so that it is now customary to impose symmetry and denote A as "SPD" for "symmetric and positive definite." I still prefer the old definition in Wachspress, 1966, but approve wholeheartedly of the use of SPD to resolve any doubt.

insight regarding best strategy for compound iteration, we ... observe that the total number of ADI iterations ... varies as

$$J\sqrt{\frac{1+\varepsilon_J}{1-\varepsilon_J}p(M^{-1}A)}. \tag{29}$$

... When Jordan's [parameter selection] is used, J is optimum when ε_J is approximately equal to 0.36. In numerical application, however one must consider relative time requirements of inner and outer iterations... It may then be best to choose J so that ε_J is an order of magnitude smaller. This may increase the total number of inner (ADI) iterations, but the overall time may be reduced significantly... A desirable feature of compound iteration is that, having decided upon strategy according to machine limitations, one may find efficient iteration parameters with negligible computation time."

The paper continued with analysis of dependence on mesh spacing as a function of normalization of A in an attempt to approach model conditions and with numerical studies comparing different normalizations. The paper concluded with the statement that "Numerical results support prediction based on theory of rapid convergence rates in the numerical solution of the diffusion equation over a rectangular domain. Further studies are contemplated, including extension to nonrectangular domains." This latter study was pursued with a few examples in my 1966 book.

3.5.3 Updated Analysis of Compound Iteration

Although much of the early analysis is still valid, developments during the past 25 years have shed new light on this approach and have led to improvements. We first consider generation of a model problem. The early studies were done with the Laplace operator as a model for the diffusion operator with diffusion coefficient $D(x, y)$. D'Yakonov proved that $p(M^{-1}A)$ is equal to the ratio of the maximum to minimum values of $D(x, y)$. This is independent of grid geometry. Thus, the number of outer iterations is independent of spacing h as $h \to 0$. Computation time per iteration increases as h^{-2} and the number of inner ADI iterations per outer iteration to achieve a fixed error reduction increases as $\log \frac{1}{h}$.

In my 1984 paper an algorithm was presented for choosing a separable model problem to solve the diffusion equation in the absence of the σ term. This requires a "best" approximation to $D(x, y)$ by the separable coefficient $D(x)D'(y)$. If one considers the approximation of $\ln D(x, y)$ by $\ln D(x) + \ln D'(y)$, one has the problem treated by Diliberto and Strauss (1951): "On the approximation of a function of several variables by a sum of functions of fewer variables." In our application we have a precise measure of merit in that now

$$p(M^{-1}A) \leq \frac{\max \frac{D(x,y)}{D(x)D'(y)}}{\min \frac{D(x,y)}{D(x)D'(y)}}. \tag{30}$$

[3]The algorithm for determining separable diffusion coefficients entails alternating improvement of $D(x)$ and $D'(y)$ until further improvement yields negligible reduction in p. The algorithm is:

1. For $i = 1,2,\ldots,m$, set $D_i = 1.0$.
2. For $j = 1,2,\ldots,n$, set

$$D'_j = \left(\max_i D_{ij} \cdot \min_i D_{ij}\right)^{\frac{1}{2}}.$$

3. For $i = 1,2,\ldots,m$, set

$$D_i = \left(\max_j \frac{D_{ij}}{D'_j} \cdot \min_j \frac{D_{ij}}{D'_j}\right)^{\frac{1}{2}}.$$

4. For $j = 1,2,\ldots,n$, set

$$D'_j = \left(\max_i \frac{D_{ij}}{D_i} \cdot \min_i \frac{D_{ij}}{D_i}\right)^{\frac{1}{2}}.$$

5. Cycle through steps 3 and 4 until values do not change appreciably. Convergence is quite rapid and high accuracy is not required. Two or three iterations often suffice.

The example given in [Wachspress, 1984] was for the pattern of diffusion coefficients in the matrix

$$D_{ij} = \begin{matrix} 9 & 25 & 1 \\ 16 & 100 & 1600. \\ 1 & 4 & 36 \end{matrix}$$

The values for D_i and D'_j obtained by two cycles of the algorithm were
$D_1 = 6.931 \quad D_2 = 28.88 \quad D_3 = 23.10$
$D'_1 = 0.465 \quad D'_2 = 12.65 \quad D'_3 = 0.237$
This resulted in

$$D_i D'_j = \begin{matrix} 1.643 & 6.845 & 5.475 \\ 87.677 & 365.332 & 292.215. \\ 3.223 & 13.429 & 10.742 \end{matrix}$$

The ratios of diffusion coefficients were then

$$\frac{D_{ij}}{D_i D'_j} = \begin{matrix} 5.478 & 3.652 & 0.183 \\ 0.183 & 0.274 & 5.475. \\ 0.310 & 0.298 & 3.351 \end{matrix}$$

[3]Al Schatz (Cornell) advised me when I was preparing work on this preconditioner for publication that he had considered a related approximation for solving finite element problems but I have not yet seen a published reference to this work. His effort was devoted more to approximating equations over nonrectangular grids by preconditioning equations over rectangular grids.

Thus, $p(M^{-1}A) = \frac{5.478}{0.183} = 29.93$ in contrast with the Laplacian model-problem value of 1600. Since the solution effort varies as the square root of p, there is a gain by a factor greater than seven through use of the best separable problem. Note that the "best" D_i and D'_j are not necessarily unique. In this example, D'_1 may vary within the interval [0.286, 0.760] without increasing p.

For the more general diffusion equation with removal σ, we first compute the separable diffusion coefficient as above and then approximate $\tau_{ij} \equiv \frac{\sigma_{ij}}{D_i D'_j}$ by $\tau_i + \tau'_j$. One scheme which has been used successfully is to approximate $\exp(\tau_{ij})$ by the product $\exp(\tau_i)\exp(\tau'_j)$, using the same algorithm as for approximating the nonseparable diffusion coefficient. Care must be taken to disallow negative removal. This can be accomplished by replacing an exponential value less than unity by unity in the algorithm.

If $\alpha > \frac{D_{ij}}{D_i D'_j} > \frac{1}{\alpha}$ and $\beta > \frac{\tau_{ij}}{\tau_i + \tau'_j} > \frac{1}{\beta}$, then $p(M^{-1}A)$ is bounded by $(\alpha + \beta)^2$. Competition between diffusion and removal is a function of the geometry and changes with mesh spacing. The removal term will have its maximum effect on eigenvectors associated with smaller eigenvalues of the matrix A. The geometric buckling of a rectangle of length X and height Y is defined as $B^2 = (\frac{\pi^2}{X^2} + \frac{\pi^2}{Y^2})$. A reasonable estimate for p is

$$p(M^{-1}A) \doteq \frac{\max\limits_{ij} \frac{B^2 D_{ij} + \sigma_{ij}}{(B^2 + \tau_i + \tau'_j) D_i D'_j}}{\min\limits_{ij} \frac{B^2 D_{ij} + \sigma_{ij}}{(B^2 + \tau_i + \tau'_j) D_i D'_j}}. \tag{31}$$

The value computed in the absence of removal is precise when there is an interior node in each region of constant $D_i D'_j$. There is an eigenvector of $M^{-1}A$ with a component of unity at each such node and zero elsewhere belonging to the eigenvalue $\frac{D_{ij}}{D_i D'_j}$. In the absence of such interior nodes, the value computed is a close upper bound on p. The value in Eq. 31 is only an estimate that can be used to assess the model problem prior to the actual iteration. In the absence of removal, precise bounds are computable for the eigenvalues of $M^{-1}A$. This facilitates use of Chebyshev extrapolation as the outer iteration. In the absence of such bounds, conjugate gradient iteration seems preferable. The cost of the additional inner products is not significant.

3.6 Interaction of Inner and Outer Iteration

Let A be the coefficient matrix of the discretized diffusion operator $-\nabla \cdot D(x, y)\nabla$ over a rectangular partitioning of a rectangle, resulting from either five-point differencing or nine-point bilinear finite elements. The vector $\mathbf{u}$ whose components are the approximations to the desired field vector at the grid nodes is obtained as the solution to the linear system

$$A\mathbf{u} = \mathbf{b}, \tag{32}$$

where $\mathbf{b}$ is a given vector. Let B be the corresponding matrix with the separable diffusion coefficient $D(x)D'(y)$, and let the model-problem matrix equation be

$$B\mathbf{v} = \mathbf{r}\,. \tag{33}$$

Let F be the SPD normalizing matrix defined in Sect. 3.1 for which the matrix splitting $B = H + V$ satisfies $HF^{-1}V - VF^{-1}H = 0$. It follows that $F^{-1}B$ commutes with $F^{-1}H$ and $F^{-1}V$. For any matrix X, define $\tilde{X} = F^{-\frac{1}{2}}\ X\ F^{-\frac{1}{2}}$. Then if we define

$$\tilde{\mathbf{v}} \equiv F^{\frac{1}{2}}\mathbf{v},\ \ \tilde{\mathbf{r}} \equiv F^{-\frac{1}{2}}\mathbf{r}\,,\ \tilde{\mathbf{b}} \equiv F^{-\frac{1}{2}}\mathbf{b}\,,\ \text{ and } \tilde{\mathbf{u}} \equiv F^{\frac{1}{2}}\mathbf{u}, \tag{34}$$

we have the transformed problem to be solved:

$$\tilde{A}\tilde{\mathbf{u}} = \tilde{\mathbf{b}}\,, \tag{35}$$

and the corresponding model problem:

$$\tilde{B}\tilde{\mathbf{v}} = \tilde{\mathbf{r}} \tag{36}$$

with $\tilde{B} = \tilde{H} + \tilde{V}$, where

$$\tilde{H}\tilde{V} - \tilde{V}\tilde{H} = F^{-\frac{1}{2}}[HF^{-1}V - VF^{-1}H]F^{-\frac{1}{2}} = 0. \tag{37}$$

Matrices $\tilde{A}$, $\tilde{B}$, $\tilde{H}$, and $\tilde{V}$ are all SPD. Let $\tilde{T}$ be the ADI iteration matrix for the symmetric normalized equations. This iteration matrix is symmetric with eigenvalues in the interval $[-\varepsilon, \varepsilon]$. The base matrix on which the outer iteration acts is

$$\tilde{W} = (I - \tilde{T})\tilde{B}^{-1}\tilde{A}, \tag{38}$$

where

$$\tilde{T}\tilde{B} - \tilde{B}\tilde{T} = 0. \tag{39}$$

A similarity transformation with $\tilde{B}^{\frac{1}{2}}$ yields

$$\tilde{W} \sim G \equiv (I - \tilde{T})\tilde{B}^{-\frac{1}{2}}\tilde{A}\tilde{B}^{-\frac{1}{2}}\,. \tag{40}$$

G is the product of two SPD matrices, $(I - \tilde{T})$ and $\tilde{B}^{-\frac{1}{2}}\tilde{A}\tilde{B}^{-\frac{1}{2}}$. Therefore, the eigenvalues of G are all real and positive and its Jordan normal form is diagonal. Let

$$b' \equiv \lambda_{\max}(\tilde{B}^{-1}\tilde{A}) = \lambda_{\max}(\tilde{B}^{-\frac{1}{2}}\tilde{A}\tilde{B}^{-\frac{1}{2}}), \tag{41}$$

and let

$$a' \equiv \lambda_{\min}(\tilde{B}^{-1}\tilde{A}) = \lambda_{\min}(\tilde{B}^{-\frac{1}{2}}\tilde{A}\tilde{B}^{-\frac{1}{2}}). \tag{42}$$

Let $b \equiv \lambda_{\max}(\tilde{W})$ and $a \equiv \lambda_{\min}(\tilde{W})$. Then

$$b \,\le\, \|G\| \,\le\, \|I - \tilde{T}\|\,\|\tilde{B}^{-\frac{1}{2}}\tilde{A}\tilde{B}^{-\frac{1}{2}}\| = (1 + \varepsilon)b' \tag{43}$$

and

$$a \geq \|G^{-1}\|^{-1} \geq [\|(I-\tilde{T})^{-1}\| \, \|\tilde{B}^{\frac{1}{2}}\tilde{A}^{-1}\tilde{B}^{\frac{1}{2}}\|]^{-1} = (1-\varepsilon)a'. \tag{44}$$

Thus, we have as rigorous bounds on the eigenvalues of $\tilde{W}$:

$$a = (1-\varepsilon)a' \text{ and } b = (1+\varepsilon)b'. \tag{45}$$

The ADI equations are not normalized with the square-root matrix. The matrix on which the outer iteration acts is now $W = (I-T)B^{-1}A$. However, a similarity transformation with $F^{\frac{1}{2}}$ reveals that $W \sim \tilde{W}$. Hence, the eigenvalues of W are all real and positive with the same bounds, and the Jordan form of W is also diagonal. Let K be the matrix of eigenvectors of W. Then $W = K\mathbf{\Lambda}K^{-1}$ where $\mathbf{\Lambda}$ is the positive diagonal matrix of eigenvalues of W. Any polynomial $P_n(W)$ can be expressed as $P_n(W) = KP_n(\mathbf{\Lambda})K^{-1}$. Therefore,

$$\|P_n(W)\| \leq \|K\|\|K^{-1}\| \max_{\lambda}|P_n(\lambda)| = \kappa(K) \max_{a\leq\lambda\leq b}|P_n(\lambda)|, \tag{46}$$

where κ is the condition number of matrix K. When Chebyshev extrapolation is used for the outer iteration with the eigenvalue bounds a and b,

$$\max_{\lambda}|P_n(\lambda)| = \left(\cosh\left[n\cosh^{-1}\left(\frac{b+a}{b-a}\right)\right]\right)^{-1}. \tag{47}$$

Thus, the norm of the error reduction after n outer iterations, with inner ADI error reduction ε each outer iteration, is bounded by

$$\sigma = \kappa\left(\cosh\left[n\cosh^{-1}\left(\frac{b+a}{b-a}\right)\right]\right)^{-1}, \tag{48}$$

where the dependence on ε occurs through $a = (1-\varepsilon)a'$ and $b = (1+\varepsilon)b'$. Rigorous bounds on b' and a' are found readily. In finite element discretization, the contribution from rectangle q to $\mathbf{x}^\top A\,\mathbf{x}$ divided by the contribution to $\mathbf{x}^\top B\,\mathbf{x}$ is $\frac{D(x,y)}{D(x)D'(y)}|_q$. Therefore, the maximum eigenvalue of $B^{-1}A$ is equal to

$$b' = \max_{x,y}\frac{D(x,y)}{D(x)D'(y)}. \tag{49}$$

Similarly,

$$a' = \min_{x,y}\frac{D(x,y)}{D(x)D'(y)}. \tag{50}$$

Let point i,j be interior to a region of constant $D(x,y)$ and $D(x)D'(y)$. Then the vector with nonzero value only at i,j is an eigenvector of $B^{-1}A$ with eigenvalue equal to $\frac{D(x,y)}{D(x)D'(y)}$. Thus, the computed bounds are actually achieved in the presence of interior nodes. The other eigenvectors are in general not easily found and have components which are mostly nonzero. The separable model problem is generated to minimize the ratio b/a. Although the ADI inner iterations required to attain a prescribed error reduction increases logarithmically with grid refinement, the number of outer iterations remains fixed. Conjugate gradient outer iteration seems appropriate in the presence of space-dependent removal terms (as in neutron diffusion problems), but when accurate eigenvalue bounds are easily found Chebyshev extrapolation may be slightly more efficient since one then avoids the need for computing two inner products per iteration.

Optimum choice of the number of inner iterations per outer may be determined in advance by minimizing the work required for a prescribed accuracy. Each inner iteration requires about the same work as the residual evaluation for the next outer iteration. Let t be the number of inners per outer and s the number of outers. Then the total work varies as $f(t) = s(1+t)$. For significant error reduction, s varies as

$$s = C\sqrt{\frac{1+\varepsilon_t}{1-\varepsilon_t}}. \tag{51}$$

Optimum strategy often requires few inners per outer so that asymptotic inner iteration convergence estimates are not valid. The AGM algorithm for $t = 2^n$ is useful in this analysis. We define

$$\theta_1 \equiv \sqrt{k'}, \tag{52.1}$$

$$\theta_m \equiv \left[\frac{2\theta_{m-1}}{1+\theta_{m-1}^2}\right]^{\frac{1}{2}}. \tag{52.2}$$

The inner iteration error reduction for t iterations is

$$\varepsilon(t) = \left(\frac{1-\theta_t}{1+\theta_t}\right)^2. \tag{53}$$

The number of outer iterations s varies as $(\theta + \frac{1}{\theta})^{1/2}$, and

$$f(t = 2^n) = C'(1+t)\left(\theta_t + \frac{1}{\theta_t}\right)^{\frac{1}{2}}. \tag{54}$$

The most efficient strategy depends on the value of k', and we examine a range of values (Table 3.1):

For most problems of interest, $k' << 0.01$ and a value close to $t = 4$ is optimum. One may compute $\varepsilon(t)$ by one of the methods described in Sect. 1.6 to optimize. For example, when $k' = 10^{-6}$, Eq. 1–54 gives

$$\begin{aligned}\varepsilon(t) &= 4\exp\left[-\frac{\pi^2 t}{\ln(4/10^{-6})}\right] \\ &= 4(0.5224)^t.\end{aligned} \tag{55}$$

For comparison with the values in the above table,

Table 3.1 Inner–outer iteration

k'	t	θ_t	$f(t)/C'$	t(opt)	ε
0.1	1	0.3162	3.73	1	0.2699
0.1	2	0.7582	6.23		
0.01	1	0.1000	6.36		
0.01	2	0.4450	4.92	2	0.1475
0.01	4	0.8619	7.11		
10^{-4}	1	0.01	20.0		
10^{-4}	2	0.1414	8.06		
10^{-4}	4	0.5317	7.76	4	0.093
10^{-4}	8	0.9105	12.8		
10^{-6}	1	0.001	63.3		
10^{-6}	2	0.0447	14.2		
10^{-6}	4	0.2991	9.54	4	0.291
10^{-6}	8	0.7410	13.0		
10^{-8}	4	0.1682	12.4	4	0.507

$$f(t) = C'\sqrt{2}(1+t)\left(\frac{1+\varepsilon}{1-\varepsilon}\right)^{\frac{1}{2}}, \tag{56}$$

and we compute $f(3)/C' \doteq 10.82$ and $f(5)/C' \doteq 9.93$. For a fair comparison we reevaluate $f(4)/C'$ with this approximation as $f(4)/C' \doteq 9.61$. This does not differ appreciably from the value of 9.54 in the table. In this case, $t = 4$ is indeed optimal.

Having established that the number of outer iterations varies as $\sqrt{p(B^{-1}A)}$ and a means for relating the number of inner iterations per outer to k', we return to the question of whether or not a nine-point model preconditioner is more efficient than a five-point model preconditioner when A is a nine-point finite element matrix. The smallest eigenvalues of B_5 and B_9 do not differ significantly. However, the largest eigenvalues differ significantly in general. One can compute these values before actually deciding on the preconditioner for a particular problem. Some insight is gained by considering the discrete Laplacian with equal mesh spacing. The B_5 and B_9 matrices have common eigenvectors in this case. However, their eigenvalues differ. The maximum absolute row sum in B_5 is 8 and the corresponding value in B_9 is 16/3. It follows that $p({B_5}^{-1}A) \doteq 1.5p({B_9}^{-1}A)$. The additional work of significance when the nine-point preconditioner is used is the recovery of the solution vector from the last iteration each cycle. This requires three flops per node. Each ADI inner iteration requires ten flops per node. Thus, if t inners are performed per outer, the additional work per outer with B_9 is by a factor of $(10t + 3)/10t$. The work ratio of nine-point to five-point iteration is then approximately equal to $(10t + 3)/(10t\sqrt{1.5}) = (10t + 3)/12.247t$. This is greater than one only when $t = 1$. When $t = 4$, which is often close to optimal, the work saving through use of the nine-point preconditioner is by a factor of approximately 1.14. One must weigh the complexity of programming a nine-point preconditioner against the gain of approximately 14 % in computation efficiency. The effect of unequal spacing should be investigated.

3.7 Cell-Centered Nodes

The five-point Laplacian discussed in Sect. 3–3.1 and the nine-point FEM discretization described in Sect. 3–3.3 are both associated with vector components computed at intersections of grid lines. An alternative cell-centered formulation also enjoys widespread application. The discretization technique is exposed by considering the operator $-\frac{\mathrm{d}}{\mathrm{d}x}D(x)\frac{\mathrm{d}}{\mathrm{d}x}$ at segment i of width h_i and diffusion coefficient D_i. The right neighboring segment is of width h_{i+1} and has diffusion coefficient D_{i+1}. The equation is integrated over segment i. The coupling between i and $i+1$ is the two-point approximation to $[-D\frac{\mathrm{d}}{\mathrm{d}x}]$ at the right end of segment i. We assume a continuous piecewise linear solution between the cell centers with joint at the segment junction. Continuity of value and current $[-D\frac{\mathrm{d}}{\mathrm{d}x}]$ at this junction yields a value there in terms of the cell-centered values of

$$u_o = \frac{D_i h_{i+1} u_i + D_{i+1} h_i u_{i+1}}{D_i h_{i+1} + D_{i+1} h_i}. \tag{57}$$

The current $[-D\frac{\mathrm{d}}{\mathrm{d}x}]$ at the junction is then approximated by

$$D_i \frac{2(u_i - u_o)}{h_i} = \frac{2 D_i D_{i+1}}{D_i h_{i+1} + D_{i+1} h_i}(u_i - u_{i+1}). \tag{58}$$

We now consider solution of Poisson's equation with a separable approximation as a preconditioner: $-\nabla \cdot D(x,y) \nabla \mathbf{u}$ approximated by $-\nabla \cdot D(x)D(y) \nabla \mathbf{u}$. We prove that when the nonseparable cell diffusion coefficient $D_{i,j}$ is approximated by the separable $D_i D_j$, the eigenvalue bounds in Eqs. 49–50 are valid. Let $\alpha_{i,j} \equiv \frac{D_{i,j}}{D_i D_j}$ and let $\alpha \le \alpha_{i,j} \le 1/\alpha$. The ratio of the true coupling between nodes i, j and $i+1, j$ and the separable approximation is

$$\begin{aligned} R(i,j) &= \alpha_{i,j}\alpha_{i+1,j} \frac{D_j(h_{i+1}D_i + h_i D_{i+1})}{h_{i+1}D_{i,j} + h_i D_{i+1,j}} \\ &= \alpha_{i,j}\alpha_{i+1,j} \frac{h_{i+1}D_i + h_i D_{i+1}}{\alpha_{i,j} h_{i+1} D_i + \alpha_{i+1,j} h_i D_{i+1}}. \end{aligned} \tag{59}$$

It follows that $R(i,j)$ is in the interval $[\alpha_{i,j}, \alpha_{i+1,j}]$. All coefficient ratios satisfy similar relationships. Hence, the eigenvalues of $B^{-1}A$ are in the interval $\alpha, 1/\alpha$ as asserted when cell-centered equations are used for the true and the model problem.

3.8 The Lyapunov Matrix Equation

Let the $n \times n$ matrix A and the SPD $n \times n$ matrix C be given. Then the Lyapunov matrix problem is to find the symmetric matrix X such that

$$AX + XA^{\top} = C. \tag{60}$$

That this Lyapunov matrix equation (and more generally the Sylvester matrix equation $AX + XB = C$, where A is of order n and B of order m) is a model ADI problem was discovered in 1982 in connection with determination of "infinitesimal scaling" impedance matrices [Hurwitz, 1984] and [Wachspress, 1988a]. Although ADI was developed for application to SPD systems with real spectra, the iteration equations do not rely on symmetry. The model condition that the component matrices commute is retained. However, the SPD condition may be relaxed to require only that the eigenvalues of the coefficient matrix lie in the positive-real half plane. Such matrices are said to be "N-stable." (The eigenvalues of a "stable" matrix are in the negative-real half plane. The "N" in N-stable is for negative and this notation implies the double negative which flips the eigenvalues into the positive-real half plane.) When A is N-stable, it is known that Eq. 51 has a unique SPD solution matrix, X. A major deterrent to use of ADI iteration for solving elliptic partial differential equations is possible loss in convergence in the absence of a convenient commuting splitting. The N-stable Lyapunov matrix problem is seen to be a model ADI problem when one recognizes that this is equivalent to a linear operator $\mathcal{A}$ mapping X into C where $\mathcal{A}$ is the sum of the commuting operators: premultiplication of X by A and postmultiplication by $A^{\top}$. Thus, commutation is inherent in the Lyapunov application.

The ADI equations applied directly to Eq. 60 are

$$X_0 = \mathbf{0}, \tag{61.1}$$

$$(A + p_j I)X_{j-\frac{1}{2}} = C - X_{j-1}(A^{\top} - p_j I), \tag{61.2}$$

$$(A + p_j I)X_j = C - X^{\top}_{j-\frac{1}{2}}(A^{\top} - p_j I), \tag{61.3}$$

with $j = 1, 2, \ldots, J$.

Matrix X is not in general symmetric after the first sweep of each iteration, but the result of the double sweep is symmetric. Each row of grid points in ADI solution of a Laplacian-type system corresponds to a column of the matrix X and each column of the Laplace grid corresponds to a row of matrix X. Equation 61.3 is actually the transpose of the conventional ADI second step. An iterative method introduced by Smith (1968) is closely related to ADI with all the p_j the same. Each of Smith's iterations effectively doubles J at the expense of three matrix multiplications.

Application of ADI iteration to N-stable Lyapunov matrix equations requires generalization of the ADI theory into the complex plane. This is described in depth in Chap. 4. The initial work concerned generalization of the elliptic-function theory and was reported in a series of papers by Ellner (nee Saltzman), Lu, and Wachspress (1986–1991). This analysis centered around embedding a given spectrum in a region bounded by a curve of the form

$$\Gamma = \{z = b\, dn[u \pm ir, k] | 0 \leq u \leq 1\}. \tag{62}$$

Such regions were denoted as "elliptic-function" regions. Additional theory relating to ADI iteration with complex spectra and methods for determining optimal ADI parameters for spectra not well represented by the elliptic-function regions used in the earlier work were reported by Starke (1989). Alternative effective parameters for rectangular spectra were developed in [Wachspress, 1991]. Subsequent analysis by Istace and Thiran (1993) applied nonlinear optimization techniques to this problem.

Popular techniques for solving Eq. 60 include the method proposed by Smith (1968) and the B–S scheme developed by Bartels & Stewart (1972). The B–S algorithm requires about $15N^3$ flops to solve for X. In many applications, neither Smith's method nor ADI iteration is competitive with B–S when applied directly to a full matrix. Even if A has a known real spectrum so that the ADI theory is precise and convergence is rapid, each iteration requires several n^3 flops. It was found that one feasible technique which makes ADI iteration competitive is to first reduce the system to banded form. ADI iterative solution when A has bandwidth $b << n$ requires only $O(bn^2)$ flops. An additional advantage of this method is that the spectrum can be determined with little increase in computation time. This facilitates choice of iteration parameters for specific spectra.

Any similarity transformation with a matrix G reduces the Lyapunov equation to

$$SZ + ZS^{\top} = D, \tag{63}$$

where

$$S = GAG^{-1} \quad Z = GXG^{\top} \quad \text{and } D = GCG^{\top}. \tag{64}$$

Once Z is found, X may be recovered from

$$X = G^{-1}ZG^{-\top}. \tag{65}$$

Reduction to diagonal form yields the solution $z_{ij} = d_{ij}/(g_{ii} + g_{jj})$, but this reduction is too costly. It is equivalent to finding all the eigenvalues and eigenvectors of matrix A. When A is symmetric, Householder reduction to tridiagonal form is efficient and robust. The spectrum is real and ADI iteration rests on theory already described. When A is not symmetric, Householder reduction may be used to transform A into upper Hessenberg form, H. ADI iteration with H is often not competitive with B–S. One may attempt to reduce H to tridiagonal form with gaussian transformations. This is a classical problem in linear algebra, known to have many pitfalls [Wilkinson, 1965]. Large multipliers often arise and these lead to rapid loss in accuracy. Several researchers addressed this problem in seeking efficient means for finding the eigenvalues of A. [Dax and Kaniel, 1981; Hare and Tang, 1989; Tang, 1988; Watkins, 1988]. Once A or H is reduced to tridiagonal form, shifted LR transformations which preserve the band structure yield the eigenvalues more efficiently than the shifted QR transformations conventionally applied to H for this purpose. Wilkinson and later researchers showed that multipliers as large as $2^{\frac{t}{3}}$ would not detract from eigenvalue accuracy for calculations performed with roundoff error of order 2^{t}. However, for solution of the Lyapunov equation, more stringent bounds are needed.

In the first numerical studies of ADI applied to the Lyapunov equation, three features were introduced. First, the gaussian reduction was applied to the Hessenberg matrix by columns, starting at the last column. Second, a recovery algorithm was applied when a large multiplier was encountered. This consisted in creating a bulge at the $(n-2,n)$ element and chasing the bulge up to the "breakdown" column [Wachspress, 1988b]. Although this often succeeded, there were situations where this did not remedy the problem. To ensure robustness, on failure of the recovery algorithm, the offending column was left intact and the algorithm was continued. This resulted in a tridiagonal system (from bounded gaussian transformations) with a few added vertical "spikes" above the diagonal. Although this was reasonably successful for the ADI iteration, it was not suitable for the eigenvalue computation since the LR iterations fill in to a full Hessenberg matrix when there are spikes. The ADI iteration lost efficiency due to insufficient spectral knowledge.

[4]A significant variant introduced in [Geist, 1989] reduced rows and columns of A sequentially from row/col 1 to row/col n. Before each row/col reduction he permuted rows and columns in an attempt to reduce the magnitude of the gaussian multipliers. Such permutations were not possible when reducing from Hessenberg form. When the row and column to be reduced are close to orthogonal large multipliers cannot be avoided. Al's program was made robust by abandoning the reduction at the point of breakdown, applying a random Householder transformation to matrix A, and restarting the reduction. With a grant from ORNL, my graduate student at the University of Tennessee (An Lu) incorporated Geist's program ATOTRI into our ADI Lyapunov solver [Geist, Lu and Wachspress, 1989]. Geist's shifted LR eigenvalue solver was then used to determine the matrix spectrum for the ADI parameter optimization.

Although most problems are solved efficiently with this procedure, the lack of robustness and the computation time expended in recovery from breakdown detract from the method. Subsequently, motivated by discussions with Al and me at ORNL, Howell (1994) handled breakdown by allowing the bandwidth to expand above the diagonal. The row reduction lagged behind the column reduction with an increase in upper-half bandwidth each time another large multiplier was encountered. In the worst case, matrix A was reduced to upper Hessenberg form by stable gaussian transformations. Howell's program BHESS [Howell and Diaa, 2005] is well suited for ADI solution of the Lyapunov equation.

The success of Geist's permutation to reduce large multipliers was puzzling since after reducing the column (which was arbitrarily reduced first) the pivot for the row is small when the row and column to be reduced are nearly orthogonal. No initial permutation can change the product of the two pivots. Large multipliers from different row/col reductions can interact to yield large norms for the composite transformation matrix and its inverse. For the Lyapunov application one should monitor the accumulated condition number of the transformation matrix.

[4]While at the University of Tennessee in Knoxville I interacted with Al Geist at Oak Ridge and awakened his interest in gaussian reduction to tridiagonal form. Our work stimulated renewed interest by several mathematicians with whom we communicated.

In 1994 I suggested a BHESS modification (described in the Howell et al. paper) which could possibly reduce interaction of large multipliers and thereby improve stability. This has been realized and is developed in Chap. 5 with application to Lyapunov and Sylvester equations.

When excessive multiplication factors do not occur, theoretical improvement over the B–S method by a factor of around two is possible with combination of reduction to banded form followed by ADI iteration. The iterative method facilitates approximate solution when solving nonlinear (Riccati) equations with Newton iteration. Each Newton iteration requires solution of a Lyapunov equation. Another beneficial property of the iterative method is that it appears to be more readily parallelizable than the B–S method, in which QR transformations consume significant computer time. The ADI iteration itself on the banded equation requires $O(bn^2)$ flops. The arithmetic associated with the similarity transformation adds up to about $7n^3$ flops.

The B–S algorithm applies to all nonsingular matrices A. The ADI iteration applies directly only when A is N-stable. When A is nonsingular but not N-stable, it is possible to transform the problem into an equivalent N-stable system [Watkins, 1988]. However, this transformation may be too expensive to justify the entire procedure. The B–S scheme seems preferable in such cases. Fortunately for the ADI alternative, many of the problems encountered are N-stable. This is evidenced by widespread use over the years of the Smith algorithm which also requires N-stability.

The minimax theory was extended for the ADI problem in analogous fashion to the polynomial approximation problem [Opfer and Schober, 1984] with Rouché's theorem replacing the Chebyshev alternating extremes property. Elliptic-function regions play the role in ADI iteration of the ellipses in polynomial approximation. The logarithms of the elliptic-function regions are close to elliptical in shape. The theory is quite definitive and yields close to optimal parameters when the spectrum can be embedded in an elliptic-function region without excessive expansion. These regions have logarithmic symmetry with respect to the real and translated imaginary axes. When such regions are not appropriate one must seek alternative parameters [Starke, 1989; Wachspress, 1991; Istace and Thiran, 1993]. Fortunately, elliptic-function regions and unions of such regions apply to many problems of concern. The ADI minimax problem is more tractable than the corresponding polynomial problem in that when the parameters are positive or appear as conjugate pairs with positive real part the spectral radius of the iteration matrix is bounded by unity.

3.9 The Sylvester Matrix Equation

The Sylvester matrix equation

$$AX + XB = C \tag{66}$$

has a unique solution X for any C when there is no combination of eigenvalues $\lambda(A)$ and $\gamma(B)$ which sum to zero. The system is then said to be nonsingular. The ADI iteration is applicable only when the sum of the real parts is positive for all combinations. Although it is possible to construct from any nonsingular system another system with the same solution for which all real part combinations are positive, this construction often involves prohibitive computation. ADI iteration does not seem to be viable for such problems.

If A and B are symmetric, solution by the method of Golub, Nash and VanLoan (1979) is quite efficient, and ADI iteration is not competitive in this case. Reduction of one of A and B to tridiagonal form and the other to diagonal form with the symmetric QR algorithm provides a robust and elegant basis for solution of the Sylvester equation. On the other hand, when A and B are not symmetric, the Householder reduction to Hessenberg form does not yield a tridiagonal matrix. The method of Golub et al. requires further reduction of only one of these Hessenberg matrices to Schur form. Nevertheless, the additional work associated with reduction to Schur form of a matrix of order n takes about $13n^3$ flops. Thus, considerable time savings may be realized through use of gaussian reduction to banded form and ADI iterative solution of the reduced equations.

Let the similarity transformations that reduce A and B to the banded matrices S and T be G and H, respectively. Then the Sylvester equation reduces to

$$SZ + ZT = F, \tag{67.1}$$

$$\text{where } S = GAG^{-1}, \tag{67.2}$$

$$T = HBH^{-1}, \tag{67.3}$$

$$F = GCH^{-1}, \tag{67.4}$$

$$\text{and } Z = GXH^{-1}. \tag{67.5}$$

The spectra for A and $A^\top$ in the Lyapunov equation were the same. Hence, parameters p_j and q_j in Eq. 62 were the same for the two steps of each iteration applied to the Lyapunov equation. Here, the spectra of A and B differ in most cases and the more general two-variable ADI theory is applicable. A generalization to complex spectra of the transformation of W.B. Jordan described in Chap. 2 will be exposed in Chap. 4. This transformation provides a basis for choice of parameters p_j and q_j.

Once A and B have been reduced to S and T of bandwidth b, one can solve the Sylvester equation by ADI iteration with $O(bnm)$ flops per iteration, where A is of order n and B is of order m. The iteration equations for the reduced system are

$$Z_0 = \mathbf{0}, \tag{68.1}$$

$$(S + p_j I_n) Z_{j-\frac{1}{2}} = F - Z_{j-1}(T - p_j I_m), \tag{68.2}$$

$$(T^\top + q_j I_m) Z_j^\top = [F - (S - q_j I_n) Z_{j-\frac{1}{2}}]^\top, \tag{68.3}$$

$$\text{for } j = 1, 2, \ldots, J$$

Let the right-hand sides in Eqs. 68.2 and 68.3 be denoted by $G_{j-\frac{1}{2}}$ and G_j. The ADI iteration arithmetic is reduced if one computes these terms recursively:

$$\text{For the first half step, } G_{\frac{1}{2}} = F \tag{69.1}$$

$$\text{and thereafter on the half steps } G_{j-\frac{1}{2}} = F + [(p_j + q_{j-1})Z_{j-1} - G_{j-1}]^\top. \tag{69.2}$$

$$\text{For the whole steps: } G_j = [F + (p_j + q_j)Z_{j-\frac{1}{2}} - G_{j-\frac{1}{2}}]^\top. \tag{69.3}$$

A rough estimate of the number of flops required to solve the Sylvester equation when $m = n$ is $21n^3$ for the Golub et al. method and $10n^3$ for the ADI method. The savings with iteration is essentially the flops associated with reduction of A or B from Hessenberg to Schur form. The iterative method uses $\frac{5}{3}(n^3 + m^3)$ flops to reduce A and B to banded form while accumulating the gaussian transformations, $nm(n + m)$ flops to transform the right-hand side, and another $nm(n + m)$ flops to recover X from Z. The estimate of $10n^3$ flops includes an allowance for the ADI iterations and verification of the approximate solution.

3.10 The Generalized Sylvester Equations

The generalized Sylvester equations may be expressed in the form

$$AX + YB = C, \tag{70}$$

$$EX - YF = G. \tag{71}$$

Matrices A and E are $n \times n$, B and F are $m \times m$, X, Y, C, and G are $n \times m$. These equations arise in solution of eigenvalue problems [Golub, Nash and VanLoan, 1979] and in control theory [Byers, 1983]. In these applications it is often true that

$$Re\lambda(E^{-1}A) + Re\lambda(BF^{-1}) > 0. \tag{72}$$

This is a stability condition which ensures existence of a unique solution to the generalized Sylvester equations. The ADI iteration equations for numerical solution of Eqs. 70–71 are

$$Y_0 = \mathbf{0}, \tag{73.1}$$

$$(A + p_j E)]X_j = C + p_j G - Y_{j-1}(B - p_j F), \tag{73.2}$$

$$(B^\top + q_j F^\top)Y_j^\top = [(C - q_j G) + (q_j E - A)X_j]^\top, \tag{73.3}$$

$$\text{for } j = 1, 2, \ldots, J.$$

These equations may be reduced to banded form. Let $S = HAE^{-1}H^{-1}$ and $T = KF^{-1}BK^{-1}$ be of bandwidth b. These matrices are computed in approximately $\frac{7}{2}(n^3 + m^3)$ flops. One also must compute $C' = HCK^{-1}$ and $G' = HGK^{-1}$. This takes $2nm(n + m)$ flops. The reduced equations are

$$Z_0 = \mathbf{0}, \tag{74.1}$$

$$(S + p_j I_n)V_j = C' + p_j G' - Z_{j-1}^{\mathrm{T}}(T - p_j I_m), \tag{74.2}$$

$$(T^{\mathrm{T}} + q_j I_m]Z_j = [C' - q_j G' - (S - q_j I_n)V_j]^{\mathrm{T}}, \tag{74.3}$$

$$\text{for } j = 1, 2, \ldots, J.$$

Note that each iteration updates both V and Z. A simple recursive relationship may be used to reduce the arithmetic in computing successive right-hand sides, denoted by L:

$$L_{\frac{1}{2}} = C' + p_1 G', \tag{75.1}$$

$$L_{j-\frac{1}{2}} = (C' + p_j G') + [(p_j + q_{j-1})Z_{j-1} - L_{j-1}]^{\mathrm{T}}, \tag{75.2}$$

$$L_j = [C' - q_j G' - L_{j-\frac{1}{2}} + (p_j + q_j)V_j]^{\mathrm{T}}. \tag{75.3}$$

This is an $O(bnm)$ algorithm with time small compared to the $O(n^3 + m^3)$ operations performed before and after solution of the equations. Matrices X and Y are recovered from V and Z with

$$X = E^{-1}H^{-1}V_J K \tag{76.1}$$

$$\text{and } Y = H^{-1}Z^{\mathrm{T}}KF^{-1}. \tag{76.2}$$

This requires another $3nm(n + m)$ flops. When $n = m$, the total arithmetic is thus around $17n^3$ flops.

3.11 The Three-Variable Laplacian-Type Problem

In Sect. 2.3 of Chap. 2 we discussed the three-variable ADI model problem and described an iteration designed primarily as a preconditioner. We will now examine this preconditioner in more detail. If we were able to use Eqs. 29.1 and 29.2 of Chap. 2 we would obtain the usual ADI preconditioning matrix, say

$$[B(t)]^{-1} = [I - M(t)]B^{-1}, \tag{77}$$

where $B = H + V + P$ is the model-problem matrix and $M(t)$ is the standard ADI iteration matrix for t double sweeps. The analysis already presented in this chapter would then apply. However, when Eqs. 2–32 are used, the preconditioner becomes

$$[B(t)]^{-1} = \left[I - \prod_{j=1}^{t} L_j \right] B^{-1}, \tag{78}$$

where L_j includes the inner ADI iteration matrix for double-sweep j of the ADI iteration. This preconditioner must be SPD for the conjugate gradient procedure to succeed. Since the L_j and B commute with one another, the preconditioner is symmetric. The norm of the ADI iteration matrix is now the spectral radius of

$$L(t) \equiv \prod_{j=1}^{t} L_j. \tag{79}$$

The spectral radius, say ε, of $L(t)$ must be less than unity for the preconditioner to be positive definite. Sufficient inner iterations must be performed to guarantee an SPD preconditioner. In Sect. 3.6 we discussed the interaction of inner ADI and outer CG iteration. We found that a value of ε of order magnitude 0.1 was reasonable and that $t = 4$ was often near optimal.

For three-variable iteration, the optimum value for t tends to be smaller than for corresponding two-variable problems. The smaller value for k'_j for the inner ADI iterations when w_j is small tends to reduce the relative efficiency as t is increased. Precise optimization of the inner ADI, the outer ADI, and the CG iteration is possible but requires evaluation of various options. This may be clarified by example.

The Dirichlet problem with Laplace's equation over a uniform grid with 100 nodes on each side yields $k' = 0.000281$ for the outer ADI iteration. The optimal parameters for $t = 4$ are $[0.000507, 0.00508, 0.05536, 0.55439]$ in the transformed space. The corresponding error reduction with Eqs. 29.1 and 29.2 of Chap. 2 is 0.0644. The inner ADI iterations for an error reduction of $\varepsilon_j < 2k' = 0.000562$ satisfy (Table 3.2).

Table 3.2 Inner iterations for a 3D problem

j	k'_j	Inner iterations
1	0.00039	8
2	0.00268	7
3	0.027	5
4	0.217	3

The spectral radius of the ADI iteration is bounded by $\varepsilon_4 = \sqrt{0.0644} = 0.2537$. The number of CG iterations is increased by iterative approximation of the model-problem inverse by a factor of 1.3. The number of mesh sweeps per CG iteration is 50. Each CG acceleration requires about the work of around three mesh sweeps. We therefore estimate the work factor as $(1.3)(53) = 68.9$.

A similar computation for $t = 2$ yields $\varepsilon_2 = 0.69$. This is achieved with 11 inner ADI iterations for a total of 24 mesh sweeps per CG iteration. The CG loss factor is now 2.33 and the estimated work factor is $(2.33)(27) = 62.9$. This is slightly better than $t = 4$.

For a two-variable computation with the same value for k', four ADI iterations per CG step would yield a work factor of 11 and two ADI iterations per CG step a factor of $(1.678)(7) = 11.75$. Although the optimum number of ADI iterations per step is now greater, these computations display an insensitivity of efficiency to the number of ADI iterations per CG step with relatively few ADI iterations being optimal.

In three-variable iteration, insufficient inner ADI iterations lead to growth in high mode $H + V$ error components. These are the oscillatory modes and their growth is similar to that associated with roundoff instability. One must not confuse this behavior with roundoff error.

Chapter 4
Complex Spectra

Abstract Recognition of Lyapunov and Sylvester matrix equations as model ADI problems stimulated generalization to complex spectra. Rouchés theorem replaces the alternating extreme property in this analysis. Complex spectra are embedded in "elliptic function regions" for which ADI iteration parameters are generated. Jordan's spectral alignment is generalized for application to Sylvester equations.

4.1 Introduction

Generalization of Chebyshev minimax theory into the complex plane has been considered by many researchers, and much of the relevant theory may be found in Smirnov (1968). A concise review of some of this theory was given by Rivlin (1980). Application to the ADI minimax problem by Starke (1989) to obtain asymptotically optimal parameters motivated recent work by Istace and Thiran (1993) in which nonlinear optimization numerical techniques were devised to determine optimal parameters. The following brief chronology was extracted from the Istace–Thiran paper: [Gonchar, 1969] characterized the general minimax problem and showed how asymptotically optimal parameters could be obtained with generalized Léja or Fejér points. [Starke, 1989] subsequently applied this theory to the ADI minimax problem. [Gutknecht, 1983] obtained a necessary Kolmogorov optimality condition for the general problem and [Ruttan, 1985] found an alternative which was refined in [Istace and Thiran, 1993]. The latter then implemented this theory for the ADI iteration problem with iterative solution of the relevant nonlinear optimization problem, starting with a complex version of an [Osborne-Watson, 1978] exchange algorithm and then switching to a Newton iteration to achieve the desired accuracy.

My analysis from 1982 to 1994 [Wachspress 1988c, 1990, 1991] was directed toward practical determination of efficient ADI iteration parameters for spectra anticipated in practice. My goal was to generalize the elliptic-function theory into the complex domain in much the same manner as the Chebyshev-polynomial theory had been generalized for polynomial approximation in the complex plane.

E. Wachspress, *The ADI Model Problem*, DOI 10.1007/978-1-4614-5122-8_4,
© Springer Science+Business Media New York 2013

Although the relevant spectra were somewhat restricted, it was found in practice that many problems could be solved expediently by embedding actual spectra in these "elliptic-function" regions. My analysis led to Starke's investigations which in turn stimulated the development of Istace and Thiran. This marriage of the more erudite and general development with my limited studies was most gratifying. One may now choose among the various approaches to find effective ADI iteration parameters for specific problems.

It was observed in Chap. 3 that my analysis of ADI iteration in the presence of complex spectra was motivated by application to solution of Lyapunov and Sylvester matrix equations. The commutation property required for application to the Dirichlet problem is not restrictive in this new application. However, complex spectra to which the ADI theory described in the previous chapters does not apply are now encountered. Complex spectra also arise in boundary-value problems containing odd order derivatives like, for example, convection diffusion equations. Theory of elliptic functions plays a prominent role in the analysis.

The coefficient matrices of the linear systems to be solved are assumed to be real and N-stable. This yields spectra in the positive-real half plane which are symmetric about the real axis. (We will designate these as PRS spectra.) It may be shown that the set of optimum parameters for any PRS spectrum must also be PRS. Unlike the corresponding polynomial approximation to zero, the rational ADI approximation is bounded in absolute value by unity for any choice of PRS parameters. It is thus possible to partition the spectral region into subregions for each of which parameters may then be selected to yield the prescribed error reduction. This was in fact the approach used by Douglas and Rachford (1956) for real spectra when ADI iteration was first introduced.

Nonsymmetric systems create other problems. Convergence of ADI iteration is retarded by deficient eigenvector spaces. Although means for handling such deficiencies may be addressed, we consider primarily problems with complete eigenvector spaces. This is assumed unless specified otherwise in the ensuing analysis. When dealing with nonsymmetric systems, the error norm reduction is not bounded by the spectral radius of the iteration matrix. If Λ is the diagonal matrix of eigenvalues of the matrix A then $A = G\Lambda G^{-1}$, where G is the matrix whose columns are the right eigenvectors of A. The condition of matrix G is $\kappa(G) = \|G\| \, \|G^{-1}\|$. If ρ is the spectral radius of an iteration matrix that commutes with A, then the norm of the error reduction is bounded by $\kappa\rho$. Nachtigal, Reddy and Trefethen (1990) considered more subtle problems associated with nonsymmetry and demonstrated for polynomial approximation to zero that one should choose parameters which minimize the spectral radius of the iteration matrix for a spectrum chosen to be slightly larger than the actual spectral region. Thus far, ADI iteration convergence with parameters based on the actual spectra has been quite satisfactory.

Just as Wilkinson's "backward error analysis" proved fruitful in studies of numerical stability, "backward spectrum analysis" has proved to be useful for ADI iteration. One chooses convenient sets of iteration parameters and determines families of spectral regions for which these sets are optimal. One then embeds a given spectrum in an approximating member of the families of spectra generated in this manner. For example, the ADI parameters determined for real spectra are also optimal for a class of complex spectra which we denote as "elliptic-function regions." One may embed a given spectral region in an elliptic-function region for which optimum parameters and the resulting error reduction are known.

Complex iteration parameters enter in conjugate pairs. By combining the two iterations with a conjugate pair, one can perform the iteration in real arithmetic with essentially no increase in computation time over that required for two iterations with real parameters.

The elliptic-function region theory has been verified with numerical solution of the Lyapunov matrix equation by ADI iteration. Some of the early studies were reported by Saltzman (1987), and later studies were reported by Lu and Wachspress (1991).

The more general development in [Starke, 1989] describes how Remez-type arguments may be applied to yield asymptotically optimal iteration parameters in similar fashion to schemes used for polynomial approximation to zero with Fejér and Léja parameters on the spectral boundary. For example, if the spectrum is bounded by a circle with real diameter the interval $[a, b]$, then the optimum parameter set is repeated use of $\sqrt{ab}$. The Fejér and Léja parameters for J iterations are J equally spaced points on the boundary. As J increases, the error reduction with these points approaches from above that obtained with the single optimum parameter repeated J times. Optimal parameters for more general regions are not found readily, but the asymptotically optimal Léja parameters can be approximated quite well by a Remez-type algorithm.

The theory for Chebyshev approximation over complex domains differs from the minimax theory for real domains. However, this complex theory applies equally as well to polynomial and rational approximation to zero. Initial analysis was for polynomials, and we lay the groundwork for the ADI iteration theory by first describing the polynomial theory.

4.2 Chebyshev Approximation with Polynomials

The general theory for rational Chebyshev approximation over complex domains is not needed for this development. We require only application of Rouché's theorem [Copson, 1935]:

ROUCHÉ'S THEOREM. *If functions f and g have no essential singularities in a region bounded by a simple closed curve on which fg is bounded and $|f|$ is everywhere greater than $|g|$, then f and $f - g$ have the same number of zeros minus poles, counting multiplicities, in the region bounded by the curve.*

This result follows directly from the Cauchy residue theorem. In our application, there are no poles within the region which contains the spectrum over which the spectral radius of the iteration matrix is to be minimized. In this section the functions are polynomials. When we treat the ADI iteration problem, the functions are rational with nonvanishing denominators over the spectral region. Suppose the iteration function f is a polynomial of degree J normalized to unity at a fixed point outside the spectral region and that it has constant absolute value H on the spectral boundary and all J of its zeros within the spectral region. If we assume the existence of a polynomial g of maximal degree J with a smaller maximum absolute value over the entire boundary, also normalized to unity at the fixed point outside the spectral region, we may apply Rouché's Theorem to arrive at a contradiction. The difference polynomial $f - g$ has J zeros inside the boundary and is zero at the normalization point. The polynomial is analytic over the spectral region and hence attains its maximum absolute value on the boundary.

One might think this result too restrictive to be of practical value since one cannot hope except in very special cases for a polynomial of this type to exist for a given spectrum. That is where the backward analysis enters. Any polynomial of degree J is optimal for a family of regions defined by its absolute level contours. The simplest example is z^J with circular contours around the origin. This leads to the important result that successive overrelaxation (SOR) applied to the usual Dirichlet-type systems with optimal extrapolation cannot be accelerated by linear combination of the SOR iterates. The eigenvalues of the SOR iteration matrix all lie on a circle. The result of J iterations is the polynomial z^J where z lies on the circle of radius equal to the spectral radius of the SOR iteration matrix. This is the unique polynomial of maximal degree J which has the least maximum absolute value over the disk bounded by the circle.

We now consider the family of regions for which Chebyshev polynomials are optimal. In my book on iterative solution of elliptic systems [Wachspress, 1966], I investigated the application of Chebyshev polynomials to spectral regions bounded by ellipses with real major axes and real normalization point. Convergence and adaptive updating of parameters was considered. There was no discussion of optimality, and the relevance of Rouché had not yet been disclosed. The first published account of application of Chebyshev polynomials to complex spectra was [Clayton, 1963] and this analysis was first applied to acceleration of Jacobi iteration in [Wrigley, 1963]. A more recent and thorough treatment of this problem was presented in [Manteuffel, 1977], who appears to be the first to discuss the relevance of Rouché's Theorem to this problem. The paper [Opfer and Schober, 1984] clarified some of the problems associated with Chebyshev approximation in the complex plane. They showed, for example, that the translated and rotated Chebyshev polynomial is not optimal for the line spectrum equal to the interval $[(1 - i), (1 + i)]$ perpendicular to the real axis, a result implicit in Manteuffel's work. (The Chebyshev parameters are asymptotically optimal for this case.) A real affine transformation normalizes a spectrum bounded by an ellipse to the region bounded by

$$\Gamma = \left\{ z = \cos(\phi + i\psi) \mid 0 \leq \phi \leq 2\pi,\ \psi = \operatorname{arc\,tanh} \frac{b}{a} \right\}, \tag{1}$$

where $2\cosh b$ is the length of the minor imaginary axis and $2\cosh a$ is the length of the major real axis of the bounding ellipse. The Chebyshev polynomial

$$C_J(z) = \cos(J \arccos z) = \cos(J\theta) \quad (2)$$

varies along Γ as

$$\cos[J(\phi + i\psi)] = \cos J\phi \cosh J\psi - i \sin J\phi \sinh J\psi. \quad (3)$$

This function has absolute value which varies between a maximum of $\cosh J\psi$ and a minimum of $\sinh J\psi$ on the boundary. The ellipse may be enclosed in a scalloped region on which $|\cos J\theta| = \cosh J\psi$. The Chebyshev polynomial is optimal over this extended region. As J increases the ratio of $\sinh J\psi$ to $\cosh J\psi$ approaches unity and the enclosing scalloped curve approaches the ellipse. Thus, the Chebyshev polynomial is asymptotically optimal.

Having established that the Chebyshev polynomials are optimal for regions which close in on an ellipse as J increases, we consider the possibility that these parameters are optimal for this ellipse for all J. I thought Tom Manteuffel had proved optimality and he responded to my query by sending me a reprint of a definitive paper by Fischer and Freund (1991) in which an error in Clayton's 1963 proof of optimality was disclosed. Fischer and Freund proved that when $J \leq 4$ the Chebyshev polynomial is optimal, but that for $J > 4$ it is not optimal when the normalization point is sufficiently close to $\cosh b$. They gave precise rules for establishing optimality. As J increases, the interval of nonoptimality decreases (as it must in view of the Rouché result). We note that as the normalization point approaches $\cosh b$ the number of iterations for significant error reduction increases. Thus, for significant error reduction J increases as the normalization point approaches $\cosh b$ and the Chebyshev polynomial approaches optimality. Moreover, the Chebyshev polynomial is optimal for all J when the ellipse is either a circle or degenerates to the real axis. Fischer and Freund were unable to detect any analytic representation of the truly optimal polynomials. In general, little is lost in practice by use of the Chebyshev polynomials. The ease with which the extrapolation parameters may be found and the associated error reduction predicted makes them well suited for this application. Their use for spectral regions other than elliptic must be examined more carefully. One must evaluate loss in convergence when the actual region is embedded in an elliptic region for selection of parameters.

There is a rather general procedure for obtaining asymptotically optimal parameters. One variant of this approach will now be outlined for a polygonal boundary. One starts with a polynomial that vanishes at each vertex. One then chooses a set of sampling points uniformly spaced on the boundary and computes the value of the iteration function at these points. If the maximum absolute value of the iteration function at sampling points between two consecutive points already selected is greater than the prescribed error reduction, then one introduces an additional parameter at the sampling point at which the function attains this maximum absolute value. This is continued until the maximum absolute value among all sampling points is less than the prescribed error reduction. These are the Léja [Léja, 1957] parameters. The scalloped boundary extension on which the constant maximum absolute value is retained defines a larger region for which the chosen parameters are optimal. This scalloped boundary approaches the actual boundary as the number of parameters is increased.

Some insight into the validity of this approximation is gained by considering the unit disk with the iteration function normalized to unity at $z = 2$. The optimum parameter is repeated use of zero, in which case the normalized iteration function for n iterations is $(\frac{z}{2})^n$ and the error reduction is 2^{-n}. The Léja points are the n roots of unity and the iteration function is $\frac{z^n-1}{2^n-1}$. Its maximum absolute value on the unit circle is $\frac{2}{2^n-1}$. As n increases this approaches $2^{(1-n)}$. Although this is twice the optimal value, the asymptotic convergence rate is the n-th root, which is $1/2$ for both the optimal and the Léja points. Thus, the Léja points are not truly optimal even asymptotically. The rate of convergence approaches the optimal rate of convergence asymptotically. The Léja points are easily generated, and the associated error reduction is found during their generation.

Relative advantages of Léja points and embedding a given spectrum in an ellipse or other region for which analytic optimization is possible must be weighed. When relatively few iterations are needed to attain the prescribed accuracy, the Léja points may be poor. On the other hand, there may be no convenient embedding of the actual region which yields efficient parameters. Then again, one must consider the advantage of having analytic error reduction bounds as a function of embedded spectral parameters. In practice, details of the spectrum are often not known and it is common to enclose known values like minimum and maximum real eigenvalues and maximum imaginary component in a conservative ellipse.

One must also consider the use of more sophisticated algorithms [Istace and Thiran] to compute optimal parameters when neither embedding in an ellipse nor using asymptotically optimal parameters is efficient. One must take care that time spent in computing parameters does not outweigh the convergence improvement derived therefrom. Significant computer programming is required for some of these schemes.

4.3 Early ADI Analysis for Complex Spectra

The earliest reported analysis of complex spectra was in [Saltzman, 1987] which applied primarily to spectra with relatively small imaginary components. A review of this analysis provides a springboard for study of more general spectra. Extensive use is made of the theory of elliptic functions and in particular Jacobi elliptic functions. My 1995 monograph drew heavily on [Abramowitz and Stegun, 1964]. The recent update of this work, [NIST Handbook of Mathematical Functions, 2010] is referenced here with the notation "N-xx.x" denoting formula or tables in Chap. 22 of the NIST handbook.

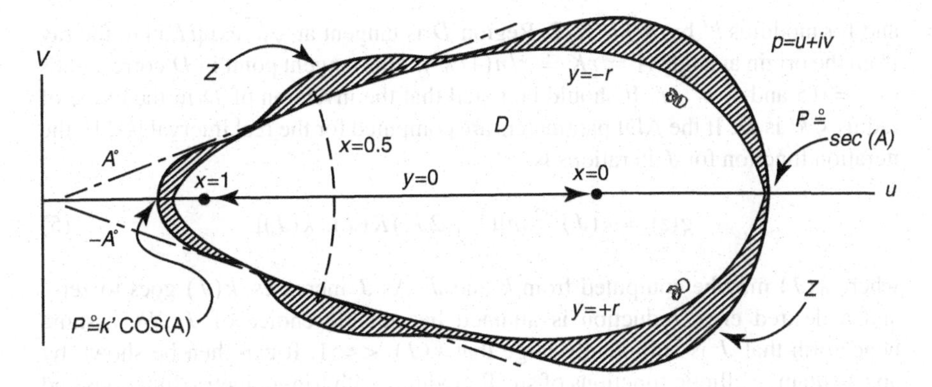

Fig. 4.1 An elliptic-function region

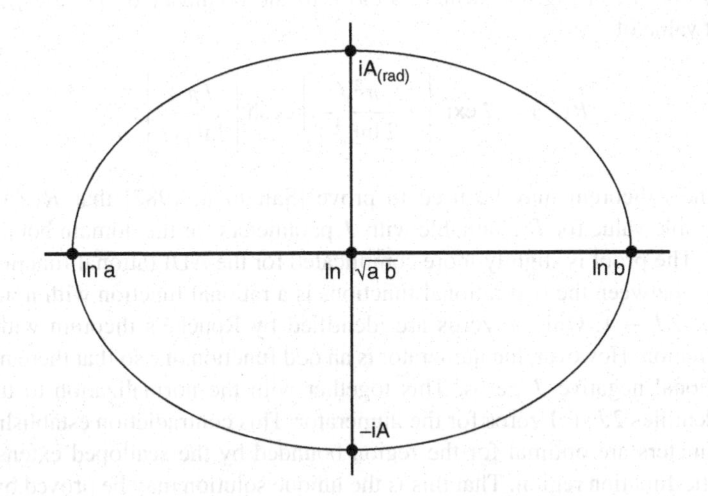

Fig. 4.2 Logarithm of an elliptic-function region

Let the eigenvalues p over which the error reduction is to be minimized for J iterations be enclosed by the elliptic-function region D:

$$D = \{p = dn(zK, k) \mid z = x + iy, \ 0 \le x \le 1 \text{ and } |y| \le r\}. \qquad (4)$$

When $1-k << 1$, D is an egg-shaped region with parameters displayed in Fig. 4.1.

The logarithm of an elliptic-function region, normalized so that the product of the endpoints of its real intercept is unity, is symmetric about both the real and imaginary axes (Fig. 4.2).

The logarithmic spectra are analogous to the elliptic spectra for polynomial approximation. The complementary modulus $k' = (1 - k^2)^{1/2}$ is less than unity. The complete elliptic integral for modulus k is approximated well by $K \doteq \ln(4/k')$

and for modulus k' by $K' \doteq \pi/2$. Region D is tangent at $\sqrt{k'}\exp[iKr]$ to the ray from the origin at angle $A = rK \doteq rln(4/k')$. This tangent point in D corresponds to $x = 0.5$ and $y = -r$. It should be noted that the inversion of D in the circle of radius $\sqrt{k'}$ is D. If the ADI parameters are computed for the real interval $[k', 1]$, the iteration function for J iterations is

$$g(z) = k(J)^{1/2} sn[(1 + 2Jz)K(J),\ k(J)], \tag{5}$$

where $k(J)$ may be computed from k and J. As J increases, $k(J)$ goes to zero, and a desired error reduction is attained by suitable choice of J. We assume henceforth that J is sufficiently large that $k(J) << 1$. It can then be shown by approximating elliptic functions of small modulus with trigonometric functions and those of modulus close to unity with hyperbolic functions that for all z on a closed curve Z (shown in Fig. 4.1) which is close to the boundary of $D, |g(z)|$ has the constant value of

$$R(Z) \doteq 2\exp\left[-\frac{\pi^2 J}{2\ln(\frac{4}{k'})}\right]\cosh\left[\frac{J\pi A}{\ln(\frac{4}{k'})}\right]. \tag{6}$$

Rouché's theorem may be used to prove [Saltzman, 1987] that $R(Z)$ is the least possible value for R attainable with J parameters for the domain bounded by curve Z. The proof is slightly more complicated for the ADI rational function. The difference between the two rational functions is a rational function with numerator of degree $2J - 1$. Only J zeros are identified by Rouché's theorem within the spectral region. However, the numerator is an odd function of z so that there must be an additional negative J zeros. This together with the normalization to unity at $z = 0$ identifies $2J + 1$ zeros for the numerator. This contradiction establishes that the parameters are optimal for the region bounded by the scalloped extension of the elliptic-function region. That this is the unique solution may be proved by more subtle arguments [Stephenson and Sundberg, 1985].

We note that $|g(z)| \leq R$ of Eq. 6 for all p in D. As J increases, Z approaches the boundary of D. Note that if we define the x-intercepts as a and b, then the modulus of the elliptic function satisfies

$$k' \doteq \frac{a}{b}\sec^2 A. \tag{7}$$

When J is large enough that cosh[·] can be approximated well by 0.5 exp[·], the value of R may be approximated by

$$R(Z) \doteq \exp\left[-\frac{\pi J(\pi - 2A)}{2\ln\frac{4}{k'}}\right]. \tag{8}$$

Thus, it is seen that as A approaches $\pi/2$, R approaches unity. This is the correct limit for eigenvalues on the imaginary axis. This form allows one to compute the loss in convergence as a function of A. For example, when $A = \pi/4$, approximately twice as many iterations are required as when A is close to zero. However, before one can use these results on problems where A is not small one must review some of the assumptions leading to Eq. 6. In particular, the value of k' in Eq. 7 is valid only for a particular range of a/b and A.

It is easily proved [Saltzman] that when the spectrum is bounded by a circle the optimum parameters are repeated use of the single value $w(j) = \sqrt{ab}$. Let $z = c + d \exp i\theta$ on the circular boundary, where $c - d > 0$. Then $a = c - d$ and $b = c + d$ and

$$\left|\frac{\sqrt{ab} - z}{\sqrt{ab} + z}\right|^2 = \frac{c - \sqrt{c^2 - d^2}}{c + \sqrt{c^2 - d^2}} = \frac{\frac{a+b}{2} - \sqrt{ab}}{\frac{a+b}{2} + \sqrt{ab}}. \tag{9}$$

The rational iteration function has constant absolute value on the circle and is, therefore, optimal. This corresponds to $k' = 1$ and $A = \arccos \frac{2\sqrt{a/b}}{1+(a/b)}$. Clearly, Eq. 7 is not valid in this case. Eq. 7 would give $k' = [(1 + a/b)/2]^2$ instead of 1. In the next section, theory will be developed for the entire range of elliptic-function domains varying from the real line to a disk and from the disk to a circle arc.

4.4 The Family of Elliptic-Function Domains

We retain the domain of Eq. 4 but drop the approximations based on $k' << 1$. The ratio of the real intercepts is obtained from Tables N-4.3 and N-6.1 as

$$\frac{a}{b} = \frac{dn[K(1+ri), k]}{dn[Kri, k]} = \frac{k'cn^2(Kr, k')}{dn^2(Kr, k')}. \tag{10}$$

By formula N-6.1, $cn^2(\text{mod } k') = (dn^2 - k^2)/k'^2$, and Eq. 10 may be solved for dn^2:

$$dn^2(Kr, k') = \frac{1 - k'^2}{1 - \frac{ak'}{b}} \tag{11}$$

We then obtain

$$cn^2(Kr, k') = \frac{\frac{a}{b}(1 - k'^2)}{k'[1 - \frac{ak'}{b}]}, \tag{12}$$

and since $sn^2 = 1 - cn^2$, we have

$$sn^2(Kr, k') = \frac{1 - \frac{a}{bk'}}{1 - \frac{ak'}{b}}. \tag{13}$$

Note that when $r = 0, k' = a/b$ is the appropriate value for this real domain and this yields $dn = 1, cn = 1$, and $sn = 0$ in the above equations.

The maximum angle is attained when $x = 1/2$ and $|y| = r$ as in the previous analysis. However, it is crucial that we not assume $k' << 1$ in evaluating this angle. As the boundary becomes more circular, k' approaches unity. Formula N-8.3 and Tables N-5.1-2 yield

$$\tan^2 A = (1-k')^2 \frac{sn^2(Kr,k')}{cn^2(Kr,k')dn^2(Kr,k')}. \tag{14}$$

Substitution of Eqs. 11–13 into Eq. 14 results in

$$\tan^2 A = \frac{(k' - \frac{a}{b})(1 - \frac{ak'}{b})}{\frac{a}{b}(1+k')^2}. \tag{15}$$

Given a domain with angle A and real intercept ratio a/b, we may solve Eq. 15 for k'. We define

$$\cos^2 B = \frac{2}{1 + \frac{1}{2}(\frac{a}{b} + \frac{b}{a})} \tag{16}$$

and

$$m = \frac{2\cos^2 A}{\cos^2 B} - 1. \tag{17}$$

If $A < B$, then $m > 1$ and we obtain from Eq. 15

$$k' = \frac{1}{m + \sqrt{m^2 - 1}} \quad \text{in } (0, 1]; \tag{18.1}$$

$$w(j) = \sqrt{\frac{ab}{k'}} dn \left[\frac{(2j-1)K}{2J}, \ k \right].$$

$$j = 1, 2, \ldots, J \tag{18.2}$$

Two limiting cases are of interest. Let $p \equiv k'b/a$. When $m >> 1$, $k' = pa/b << 1$ and from Eq. 15, $\tan^2 A \doteq p - 1$ or $p \doteq \sec^2 A$ as in Eq. 7.

Next let $m = 1$, the smallest value for which k' remains real. In this limit $k' = 1$ and Eq. 15 yields $\tan A = \frac{(1-\frac{a}{b})}{2\sqrt{a/b}}$. Hence, $\cos A = \frac{2\sqrt{a/b}}{(1+\frac{a}{b})}$ as was previously derived for the disk. It is thus established that these new relationships provide a transition between the real line and the disk. It should be noted that in this limit $Kr \to K'$ which becomes infinite at $k' = 1$ and that the elliptic-function region does approach a disk rather than a point.

To illustrate a spectrum in the transition region, let $a/b = 0.1$ and $A = 45°$. Then $m = 2.025$ and $k' = 0.264$. Note that $(a/b)\sec^2 A = 0.2$. The larger value of k' here reflects the contraction of the parameters toward $\sqrt{ab}$ as the domain moves from the real line to the disk. When $m \geq 1$ the optimum parameters are real. If $m < 1$ all the parameters lie on an arc of the circle of radius $\sqrt{ab}$.

We preempt analysis with elliptic functions of complex moduli by defining a dual spectrum. To motivate the dual spectrum technique, we consider optimum parameters for the spectrum consisting of the arc of the unit circle between $-A$ and $+A$. The folding $z' = \frac{(z+\frac{1}{z})}{2}$ transforms the arc into the real interval $[\cos A, 1]$. The dual interval $[a, 1/a]$ folds into $[1, \sec A]$ when $a = \tan(\pi/4 - A/2)$. Hence, if we first compute the optimum parameters over $[a, 1/a]$ for J iterations, these parameters will fold into parameters over the interval $[1, \sec A]$ which give the proper Chebyshev alternating extremes property. The reciprocal of these parameters will retain this property over the interval $[\cos A, 1]$. The inverse transformation back to the arc will then yield the optimum parameters over the arc! When J is odd, the real parameter at angle $A(j) = 0$ is used only once and all the other parameters on $[\cos A, 1]$ transform back into the $\pm$ angles on the arc.

The recipe for computing the optimum $A(j)$ for $j = 1, \ldots, J$ derived on this basis is

$$k' = \tan^2(\pi/4 - A/2), \tag{19.1}$$

$$z(j) = \frac{1}{\sqrt{k'}} dn\left[\frac{(2j-1)K}{2J}, k\right], \tag{19.2}$$

$$w(j) = \frac{1}{2}\left[z(j) + \frac{1}{z(j)}\right],$$

$$j = 1, 2, \ldots, \text{ integer part of } \frac{1+J}{2}, \tag{19.3}$$

$$A(2j-1) = \arccos\frac{1}{w(j)}, \tag{19.4}$$

$$A(2j) = -A(2j-1), \tag{19.5}$$

$$w(j) = \exp[iA(j)]. \tag{19.6}$$

When J is odd, the value $A[(1+J)/2] = 0$ is not repeated. This technique generalizes to elliptic-function spectra. The actual and dual spectra fold into reciprocal spectra with m' for the dual spectrum > 1 when m for the actual spectrum is < 1. The algebra is not trivial, but the resulting equations are easily verified. The duality relationships are remarkable. They highlight an elegant application of classical analysis to a crucial problem of numerical analysis.

The elliptic spectrum is defined by the triplet $\{a, b, A\}$. The dual elliptic spectrum is defined by the triplet $\{a', 1/a', A'\}$, with $A' = B$ of Eq. 16 and

$$a' = \tan\left(\frac{\pi}{4} - \frac{A}{2}\right). \tag{20}$$

Substituting this value for a' into Eq. 16, we find that

$$B' = A \tag{21}$$

and therefore

$$m' = \frac{2\cos^2 B}{\cos^2 A} - 1 \tag{22}$$

must be greater than 1 when $m < 1$. We use m' in place of m in Eq. 18.1 and compute the optimum real parameters $\{w'(j)\}$ for the dual problem. The corresponding parameters for the actual elliptic spectrum may then be computed from

$$\cos A(j) = \frac{2}{w'(j) + \frac{1}{w'(j)}}$$

$$\text{for } j = 1, 2, \ldots, \text{integer part of } \frac{1+J}{2}, \tag{23}$$

$$w(2j-1) = \sqrt{ab}\ \exp[iA(j)], \tag{24}$$

and

$$w(2j) = \sqrt{ab}\ \exp[-iA(j)]. \tag{25}$$

When J is odd, Eq. 25 is dropped for $2j = J + 1$ and the last value computed by Eq. 24 is $w(J) = \sqrt{ab}$.

This is illustrated with $\{a, b, A\} = \{0.1, 1.0, 60°\}$. We compute $\cos^2 B = 2/(1.0 + 10.1/2) = 0.3306$ and since this is greater than $\cos^2 60° = 0.25, m = 0.5125 < 1$. For the dual problem, $m' = 2(.3306)/0.25 - 1.0 = 1.645$ and $k' = 0.3389$. For $J = 2$ we compute $w'(1) = 0.685114$ and $w'(2) = 1.45961$, so $2/[w'(1) + 1/w'(1)] = 0.93252$ and $A(1) = \arccos(0.93252) = 0.3695$ rad. The optimum parameters are therefore $w(1) = 0.316\exp(0.3695i)$ and $w(2) = 0.316\exp(-0.3695i)$.

The range of elliptic-function domains for which we have now developed a theory for computing nearly optimum parameters is displayed in Fig. 4.3.

Rouché's theorem establishes that our elliptic-function parameters are optimal for a scalloped region which approaches the elliptic-function region as J increases. Thus, the parameters are "nearly optimal" in the same sense that Chebyshev parameters are nearly optimal for polynomial extrapolation. We now consider the possibility that the elliptic-function parameters may be optimal even when J is small. We have already described how Chebyshev polynomials are not optimal for polynomial approximation to zero over elliptic spectra in certain cases. Does this result carry over into rational approximation with elliptic functions over elliptic domains? Although it may be true that there are cases where the elliptic functions are suboptimal, it is easily shown that the polynomial result does not apply.

Theorem 12 (Optimality of elliptic parameters when J=2^n). *When $J = 2^n$, the elliptic-function parameters are optimal over elliptic-function spectral domains.*

Proof. The algorithm for obtaining optimal parameters over a real spectral interval when $J = 2^n$ applies to elliptic-function spectra. Successive Landen transformations (N-7) and renormalizations reduce the number of parameters to one, and this optimal parameter is the geometric mean of the endpoints of the real intercept. The back transformations yield the usual elliptic-function parameters. □

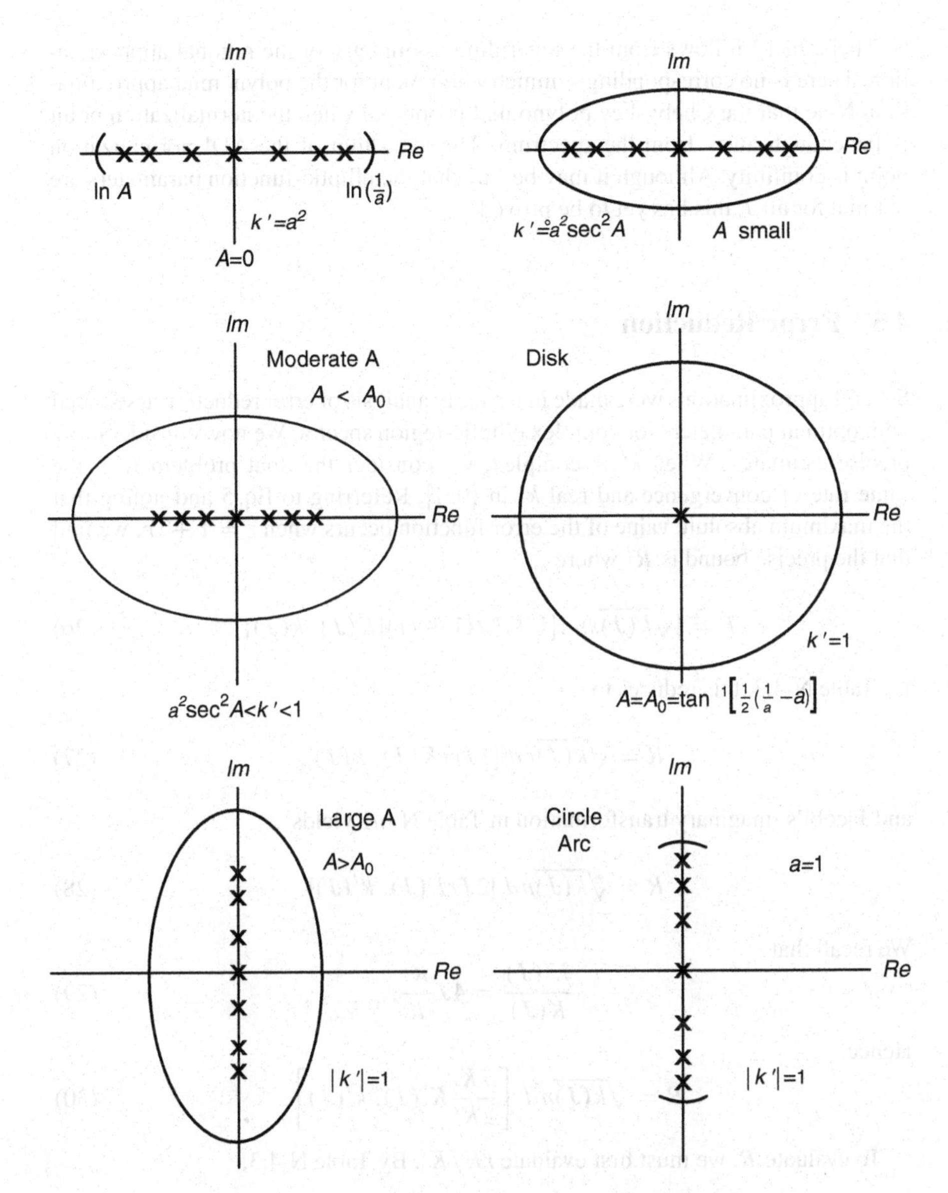

Fig. 4.3 Log of elliptic-function spectra ($0 \leq A \leq \pi/2$) x = locations of optimum iteration parameters for $J = 7$

Theorem 12 follows from the logarithmic symmetry of the rational approximation. There is no corresponding symmetry argument for the polynomial approximation. Note that the Chebyshev polynomial is optimal when the normalization point is far enough away from the spectrum. The logarithm of the ADI normalization point is at infinity. Although it may be true that the elliptic-function parameters are optimal for all J, this has yet to be proved.

4.5 Error Reduction

Several approximations were made in the early analysis of error reduction associated with optimal parameters for complex elliptic-region spectra. We now consider more precise estimates. When k' is complex, we consider the dual problem with the same rate of convergence and real k' in (0, 1]. Referring to Eq. 5 and noting that the maximum absolute value of the error function occurs when $z = 1 + ir$, we find that the precise bound is R^2 where

$$R = |\sqrt{k(J)}sn\,\{[1 + 2J(1 + ri)]K(J), k(J)\}|. \tag{26}$$

By Table N-4.3, this reduces to

$$R = \sqrt{k(J)}cd\,[2JriK(J), k(J)], \tag{27}$$

and Jacobi's imaginary transformation in Table N-6.1 yields

$$R = \sqrt{k(J)}nd\,[2JrK(J), k'(J)]. \tag{28}$$

We recall that

$$\frac{K'(J)}{K(J)} = 4J\frac{K'}{K}. \tag{29}$$

Hence,

$$R = \sqrt{k(J)}nd\left[\frac{rK}{2K'}K'(J), k'(J)\right]. \tag{30}$$

To evaluate R, we must first evaluate rK/K'. By Table N-4.3,

$$sn\,[(K' - rK), k'] \equiv cd\,(rK, k') \equiv \frac{cn\,(rK, k')}{dn\,(rK, k')}. \tag{31}$$

By Eqs. 11 and 12,

$$cd\,(rK, k') = \sqrt{\frac{a}{bk'}}, \tag{32}$$

and we may compute $K' - rK$ with the AGM$(1, k)$ algorithm in Sect. 1.6, starting with $\phi_0 = \arcsin\sqrt{\frac{a}{bk'}}$ and using Eq. 51 of Chap. 1to compute ϕ_N. We have

$$K' - rK = \frac{\phi_N}{2^N a'_N} \text{ with } \phi \text{ in radians}$$

$$= \frac{\phi_N \pi}{2^N (180) a'_N} \text{ with } \phi \text{ in degrees.}$$

Since $K' = \pi/2a'_N$, we obtain when ϕ is expressed in degrees

$$v \equiv 1 - \frac{rK}{K'} = \frac{\phi_N}{90 \cdot 2^N}. \tag{33}$$

We now define

$$y \equiv \frac{1}{2}(1 - v) = \frac{rK}{2K'} \tag{34}$$

and

$$w \equiv q(J) = q^{4J}. \tag{35}$$

We may compute the nd-function in Eq. 30 with the AGM algorithm to any desired accuracy. However, there is a simple approximation to R which seems adequate for virtually all applications. We substitute the approximation to the dn-function given in Eq. 56 of Chap. 1 into Eq. 30 to obtain

$$R \doteq w^{\frac{1-2y}{4}} \frac{1 + w^y + w^{2-y}}{1 + w^{1-y} + w^{1+y}}. \tag{36}$$

The value for q may be approximated as described in Sect. 1.6. Equations 50.1–50.3 of Chap. 1 are often suitable for this purpose. We recall that the error reduction is equal to R^2. As J increases, R approaches q^{vJ} and the asymptotic rate of convergence is $\rho_\infty = q^{2v}$.

The procedure will now be illustrated for the spectrum $\{a, b, A\} = \{0.1, 1.0, 45°\}$, for which we have already determined that $k' = 0.26414$. We compute $k = 0.96448$ and $\phi_0 = \arcsin(\sqrt{\frac{a}{bk'}}) = 37.973°$. The AGM algorithm converges to five digit accuracy after two steps:

Table 4.1 An AGM table

n	$a(n)$	$b(n)$	$c(n)$	$b(n)/a(n)$	ϕ_n
0	1.0	0.96448	0.26414	0.96448	37.973°
1	0.98224	0.98208	0.01776	0.999837	74.946°
2	0.98216	0.98216	0.00008	1.0000	149.89°

Thus, $v = 149.89/360 = 0.416361$ and $y = (1 - v)/2 = 0.29182$. The approximations in Eqs. 50 of Chap. 1 yield

$$z \doteq \frac{1}{2}\frac{1-\sqrt{k}}{1+\sqrt{k}} = 0.00452,$$

$$q' \doteq z(1+z^4) \doteq 0.00452$$

and Eq. 48 of Chap. 1 yields $q = \exp[\frac{\pi^2}{\ln q'}] = 0.16074$. Setting $w = q^{4J}$, we compute R with Eq. 36 for $J = 1, 2, 4$ and compare with truth determined with the optimal parameters and the actual rational function evaluated at the point $x = 0.1$.

Table 4.2 Convergence rate estimates

J	1	2	4
R(Truth)	0.5195	0.2213	0.04763
R(Eq. 36)	0.4987	0.2208	0.04763
q^{vJ}	0.46715	0.2182	0.04763

The last row indicates how rapidly the asymptotic convergence rate is attained when k' is not close to zero. This row was computed with $q^v = 0.46715$ (Table 4.2).

It is instructive to examine some limiting cases. When $\frac{a}{b} << 1$ and angle A is small, we set $k' = p\frac{a}{b}$ in Eq. 15 and find that $p = \sec^2 A$ as in the early result in Eq. 7. It follows that $\sqrt{\frac{a}{bk'}} = \cos A$. Moreover, when $k' << 1$, $sn(z, k')$ may be approximated (N-10.4) by $\sin(z)$. From Eqs. 31 and 32, we have

$$sn\,[(K' - rK), k'] = \sqrt{\frac{a}{bk'}} \doteq \cos A \doteq \sin(K' - rK). \tag{37}$$

In this case, $K' \doteq \frac{\pi}{2}$ and it follows that

$$rK \doteq A. \tag{38}$$

The asymptotic convergence rate is now obtained by allowing J to increase until $k(J) << 1$. By N-10.9, $nd(z) < \cosh(z) < e^z$. Equation 28 yields

$$R(J) < 2\exp\left[-\frac{\pi JK'}{K}\left(1 - \frac{2A}{\pi}\right)\right]. \tag{39}$$

Although K' remains close to $\pi/2$ as A increases slightly from 0, the value for K decreases from $\ln\frac{4b}{a}$ to $\ln\frac{4b\cos^2 A}{a}$. Thus, as A increases the number of iterations required for prescribed error reduction increases by a factor of

$$\frac{J(A)}{J(0)} \doteq \frac{1 + \frac{2\ln(\cos A)}{\ln(4/k')}}{1 - \frac{2A}{\pi}}. \tag{40}$$

For small k', the decrease in K is insignificant and the ratio is approximately $(1 - 2A/\pi)^{-1}$. Even though the approximation applies to small A, we observe that an increase by a factor of two occurs when A is near $\pi/4$. Relatively large complex components do not seem to be as detrimental to ADI iteration as corresponding components are to polynomial approximation.

We may also consider convergence as the elliptic-function region approaches a disk, in which case k' approaches unity and both K and $K(J)$ are close to $\pi/2$. By N-10.7, $sn(z, k')$ may be approximated well by $\tanh(z)$, and we obtain the approximation

$$K' - rK = \operatorname{arc\,tanh} \sqrt{\frac{a}{bk'}}. \tag{41}$$

Now even for small J, $k'(J)$ is close to unity and

$$nd\,[2rJK(J), k'(J)] \doteq \cosh[2rJK(J)] \doteq \cosh(rJ\pi). \tag{42}$$

For J sufficiently large,

$$\sqrt{k(J)} \doteq 2\exp\left(-\frac{\pi JK'}{K}\right) \tag{43}$$

and the approximation $\cosh[\cdot] \doteq \exp[\cdot]/2$ yields

$$R(J) \doteq \exp\left[-\frac{\pi J}{K}(K' - rK)\right] \doteq \exp[-2J(K' - rK)]. \tag{44}$$

Substituting Eq. 41 into Eq. 44 and using the identity

$$\operatorname{arc\,tanh} u = \ln\left(\frac{1+u}{1-u}\right)^{\frac{1}{2}}, \tag{45}$$

we obtain

$$R(J) = \left[\frac{1 - \sqrt{\frac{a}{bk'}}}{1 + \sqrt{\frac{a}{bk'}}}\right]^J. \tag{46}$$

The example with $(a, b, A) = (0.1, 1.0, 45°)$ is not far from a disk, and Eq. 46 yields an approximate asymptotic convergence rate of $\rho_\infty = 0.23816$ and $R(J) \doteq (0.488)^J$, which may be compared with the correct asymptotic value of $(0.46715)^J$.

When $k' = 1$, the region is a disk and $R(J)$ attains the correct limit of

$$R(J) = \left[\frac{1 - \sqrt{\frac{a}{b}}}{1 + \sqrt{\frac{a}{b}}}\right]^J. \tag{47}$$

Added in 2012: The maximum for $|R|$ on the boundary of the elliptic-function region occurs at $w = 1 + ir$ as in Eq. 26. The minimum absolute value on the boundary occurs when $w = 0.5 + ir$, in which case Table N-4.3 yields $R(u = 0.5) = \sqrt{k(J)}sc[2JrK(J), k'(J)]$. By Rouché's theorem, the optimum parameters for the elliptic-function region cannot result in a lower value for $|R|$. For significant error reduction $k(J)$ is small and $k'(J)$ is close to unity. The elliptic-function region may be approximated by hyperbolic functions (cf. N-10.7-9) and the ratio of the minimum to maximum over the boundary is $\tanh[2rJK(J)]$. Here $K(J) \doteq \pi/2$ and the ratio is $\tanh(rJ\pi)$. When $k' << 1$, by Eq. 38 $r = \frac{A}{K}$ and the greatest loss in error reduction through use of the elliptic rather than optimum parameters is bounded by

$$L \equiv \left|\frac{R^2(u = .5)}{R^2(u = 1)}\right| = \tanh^2\left(\frac{\pi A J}{K}\right).$$

By Eq. 8, the error reduction is $\varepsilon = \exp[-\frac{\pi J(\pi-2A)}{K}]$, from which we find that

$$\frac{\pi J}{K} = \frac{\ln\frac{1}{\varepsilon}}{\pi(1 - \frac{2A}{\pi})}.$$

When A is small this analysis is not relevant since $A = 0$ and $L = 0$. However, the elliptic parameters are optimum for this real spectrum. When A is sufficiently large we define

$$\zeta = \frac{\frac{2A}{\pi}}{(1 - \frac{2A}{\pi})}$$

and observe that for significant error reduction $L = 1-2\varepsilon^{\zeta}$. For example, when $A = \pi/6, \zeta = 1/2$ and $L = 1 - 2\sqrt{\varepsilon}$ are close to unity. The elliptic parameters are close enough to optimum to justify embedding an actual spectrum in an elliptic-function region whenever feasible without undue enlargement of the spectrum. Algorithms for generating truly optimum parameters cannot yield significant improvement for such regions and may be quite complicated and time consuming. In practice the spectrum may not be known well enough to preclude embedding in a conservative elliptic-function region.

4.6 The Two-Variable Problem

4.6.1 Generalized Spectral Alignment

A need for treating complex spectra first arose with the discovery that the Lyapunov matrix equation $AX + XA^T = C$ is a model ADI problem when matrix C is SPD and the eigenvalues of matrix A are in the positive-real half plane. For this application the two spectral regions are the same. We have just exposed theory for treating this case. The Sylvester matrix equation is $AX + XB = C$, where A is a given $m \times m$ real matrix, B is a given $n \times n$ real matrix, C is a given $m \times n$ real matrix, and the $m \times n$ real matrix X is to be determined. This is an ADI model problem when the eigenvalues λ and γ of matrices A and B satisfy $\min[Re\lambda(A)] + \min[Re\gamma(B)]>0$. In general, the eigenvalues of A and B are complex and the spectra of these matrices may differ widely. Our goal is to generalize the real transformation in order to align two complex spectra of this type. We first apply a WBJ transformation to align the real intercepts at $[k', 1]$ and normalize to $[\sqrt{k'}, 1/\sqrt{k'}] \equiv [e, 1/e]$. We may then treat eigenvalues subtending large angles (e.g., greater than 1 rad) discretely, thus removing them from the spectra to be aligned. The maximum angles of the remaining spectra may then be found as θ_1 and θ_2 in $S_1 = (e, 1/e, \theta_1)$ and $S_2 = (e, 1/e, \theta_2)$. If optimal parameters can be found for these spectra, they may be transformed back to parameters for the actual spectra. By Eqs. 49.1 and 49.2, further alignment is needed when the angles differ.

The optimum parameters for J iterations over spectrum $S = (e, 1/e, \theta)$ are given in Eq. 18.2 as

$$w_j = \frac{1}{\sqrt{k'}} dn\left[\frac{2j-1}{2J}K, k\right], \quad j = 1, 2, \ldots, J, \tag{48}$$

where k' depends on e and θ as shown in Eqs. 16–18.1. Useful parameters for this analysis are defined as

$$f_e \equiv \frac{1}{2}\left(e + \frac{1}{e}\right), \tag{49.1}$$

$$\zeta \equiv f_e \cos\theta. \tag{49.2}$$

Then it is easily shown that k' may be computed with Eqs. 18.1 and 18.2 from ζ, when one observes that

$$m = 2\zeta^2 - 1, \tag{49.3}$$

$$k' = \frac{1}{m + \sqrt{m^2 - 1}}. \tag{49.4}$$

Having aligned the real intercepts of our two spectra, we observe that when $\theta_1 \neq \theta_2$ they do not share the same optimum parameters. The ratio $\cos\theta_1/\cos\theta_2$ is a measure of the disparity of the two spectra. If this ratio is close to unity, we may choose parameters for the larger of the two angles. Each parameter w_j may then be transformed back to yield the corresponding values for p_j and q_j. In general, the two angles will differ. We seek another transformation to align the spectra. To this end, we first establish a relationship between s and t such that there is a Jordan-type transformation which maps $S_1 = (e, 1/e, \theta_1)$ onto $S_1(s) = (s, 1/s, \psi_1)$ and $S_2 = (e, 1/e, \theta_2)$ onto $S_2(t) = (t, 1/t, \psi_2)$. In an attempt to accomplish this objective, we repeat the analysis in Chap. 2 with a few simple modifications.

4.6.2 A One-Parameter Family of Spectral Pairs

We first define

$$K = \begin{bmatrix} s & 0 \\ 0 & 1/s \end{bmatrix}, \quad L = \begin{bmatrix} t & 0 \\ 0 & 1/t \end{bmatrix}, \quad A = \begin{bmatrix} 1 & -e \\ 1 & -1/e \end{bmatrix}, \text{ and } F = \begin{bmatrix} 1 & e \\ 1 & 1/e \end{bmatrix}. \tag{50}$$

Then the matrix C in Eqs. 2–10 is replaced by

$$C = \begin{bmatrix} KA & A \\ LF & -F \end{bmatrix} = \begin{bmatrix} KAF^{-1} & 0 \\ L & -I \end{bmatrix} \begin{bmatrix} F & FA^{-1}K^{-1}A \\ 0 & (F + LFA^{-1}K^{-1}A) \end{bmatrix}. \tag{51}$$

Now we define $G \equiv FA^{-1}K + LFA^{-1}$ and note that G can be singular only when

$$\det \begin{bmatrix} (s+t)(e+1/e) & -2e(t+1/s) \\ 2(s+1/t)/e & -(e+1/e)(1/s+1/t) \end{bmatrix} = 0. \tag{52}$$

We determine that Eq. 52 is satisfied when

$$t = \frac{1 - sf_e}{f_e - s}. \tag{53}$$

Algebra identical to that used in Chap. 2 now yields corresponding values for the transformation parameters of

$$\alpha' = \delta' = 1 - es, \qquad \beta' = \gamma' = e - s, \tag{54}$$

Since in this application $\alpha' > 0$, we may normalize to $\alpha = 1$. We identify S_1 as the spectrum with the smaller angle. Then $s > e$ and we define the normalized positive $\beta \equiv -\beta'/(1 - es) = (s - e)/(1 - es)$ to obtain the transformations

$$z_1 = \frac{w_1 - \beta}{1 - \beta w_1} \text{ and } z_2 = \frac{w_2 + \beta}{1 + \beta w_2}. \tag{55}$$

These transformations map the unit circle into the unit circle.

We now seek a value for β such that

$$\zeta_1 \equiv f_s \cos \psi_1 \tag{56}$$

and

$$\zeta_2 \equiv f_t \cos \psi_2 \tag{57}$$

with $\zeta_1 = \zeta_2$. If this is possible, then the transformed spectra are aligned and optimal parameters for these spectra may be transformed back.

4.6.3 Transformation from [e/1/e] to [s,1/s] and [t,1/t]

The inverses of the transformations in Eq. 55 are

$$w_1 = \frac{z_1 + \beta}{1 + \beta z_1} \quad \text{on } S_1 \quad \textit{and} \quad w_2 = \frac{z_2 - \beta}{1 - \beta z_2} \quad \text{on } S_2. \tag{58}$$

We recall that these transformations leave the unit circle invariant. We now prove that the transformed spectra remain elliptic-function regions.

Theorem 13. *If*

$$z(u + iv) \equiv \frac{1}{\sqrt{k_0}} dn\,[(u + iv)K_0, k_0] \; \textit{and}$$

$$w(u + iv) \equiv \frac{z + \beta}{1 + \beta z}, \; \textit{then}$$

$$w(u + iv) = f(u + iv) \equiv \frac{1}{\sqrt{k}} dn\,[(u + iv)K, k],$$

where k is uniquely determined by β and k_0.

Proof. We first observe that $f(1/2) = w(1/2) = 1$. The real and imaginary periods of f and w are the same. If we can choose k so that they have the same zeros and poles, the functions are the same. We have $w = 0$ when $u = 1$ and $v = r_0$, where r_0 is determined from $dn[(1 + ir_0)K_0, k_0] = nd(ir_0K_0, k_0) = cd(r_0K_0, k'_0) = -\beta\sqrt{k_0}$. If $\beta = 0, ir_0 = \tau_0$ and $r_0 = K'_0/K_0$. If $\beta > 0, ir_0 > \tau_0$, and if $\beta < 0, ir_0 < \tau_0$. The value for k is chosen so that $dn[(1 + ir_0)K, k] = 0$. This is true when $ir_0 = \tau$. Now $dn(ir_0K, k) = dn(iK', k) = \infty$ so that ir_0 is a pole of $f(u + iv)$. We now note that

$$dn(ir_0K_0, k_0) = \frac{k_0}{dn[(1 + ir_0)K_0, k_0]} = -\frac{\sqrt{k_0}}{\beta},$$

so that $z(ir_0) = -1/\beta$ and

$$w(ir_0) = \frac{\beta - \frac{1}{\beta}}{1 - \beta\frac{1}{\beta}} = \infty.$$

Thus w and f have the same poles. Both functions are elliptic, their ratio is unity at one point, they have the same periods, and they have the same zeros and poles. It follows that they are equal. □

Having established that the transformed region remains elliptic, we need not determine r_0 to evaluate k. We need only compute the angle subtended by the unit circle and compute k' from Eq. 49.

Inspection of Eq. 58 reveals that

$$p \equiv (1 + st)/(s + t) \tag{59}$$

is an invariant of these transformations.

The transformation to real intervals $[s, 1/s]$ and $[t, 1/t]$ is accomplished with

$$\beta = (s - e)/(1 - se). \tag{60}$$

The angles for the transformed spectra are determined as

$$\theta_s = \arccos\left[\frac{(1 + \beta^2)\cos\theta_1 + 2\beta}{(1 + \beta^2) + 2\beta\cos\theta_1}\right] \tag{61}$$

and

$$\theta_t = \arccos\left[\frac{(1 + \beta^2)\cos\theta_2 - 2\beta}{(1 + \beta^2) - 2\beta\cos\theta_2}\right]. \tag{62}$$

We now define

$$f_s = \frac{1}{2}\left(s + \frac{1}{s}\right) \quad \text{and} \quad f_t = \frac{1}{2}\left(t + \frac{1}{t}\right). \tag{63}$$

Then the values for ζ are

$$\zeta_s = f_s \cos\theta_s \quad \text{and} \quad \zeta_t = f_t \cos\theta_t. \tag{64}$$

The spectra are aligned when $\zeta_s = \zeta_t$. We associate S_1 with the spectrum for which ζ is larger and seek a value for β which will yield $\zeta_s = \zeta_t$.

4.6.4 *Alignment When $\zeta_s > 1$ and $\zeta_t > 1$*

We attempt alignment by increasing s. Invariance of p in Eq. 59 establishes that t decreases according to

$$t = (1 - sp)/(p - s). \tag{65}$$

Thus, as s approaches $1/p$, t approaches zero. It appears that we may decrease t until the spectra are aligned, but this is not always possible. Since $\beta > 0$, the numerator in the expression for θ_t in Eq. 62 decreases as s increases so that the decrease in $\cos\theta_t$ can dominate the increase in f_t. We first show that the spectra may be aligned when $\zeta_1 > \zeta_2 > 1$

Theorem 14. *If $\zeta_2 > 1$, then θ_t is bounded away from $\pi/2$ for $0 < t < t_0$.*

Proof. As s increases, θ_t increases to its maximum at $s = 1/p$ at which point $\beta = t_0$. Substituting this value for β into Eq. 62, we find that

$$\cos\theta_{t=0} = \frac{(1+t_0^2)\cos\theta_2 - 2t_0}{(1+t_0^2) - 2t_0\cos\theta_2} = \frac{\zeta_2\cos\theta_2 - 1}{\zeta_2 - \cos\theta_2},$$

which is in the interval (0, 1]. Hence,

$$\theta_t < \theta_{t=0} < \frac{\pi}{2}, \qquad 0 < t < t_0. \tag{66}$$

□

It follows that as s is increased, $\zeta_s = f_s\cos\theta_s < f_s < f_{s0}$ and $\zeta_t = f_t\cos\theta_t > f_t\cos\theta_{t=0}$. Thus, as t approaches zero f_t increases until at some point in $(0, t_0)$ $\zeta_t = \zeta_s$. The spectra can always be aligned when ζ_1 and ζ_2 are both greater than one. Let

$$\tau \equiv \frac{\beta + \frac{1}{\beta}}{2}. \tag{67}$$

Then

$$\begin{aligned} s + \frac{1}{s} &= \frac{e+\beta}{1+e\beta} + \frac{1+e\beta}{e+\beta} \\ &= \frac{(1+e^2)(1+\beta^2)+4e\beta}{(1+e^2)\beta + e(1+\beta^2)} \\ &= \frac{2(f_e\tau+1)}{\tau+f_e}. \end{aligned}$$

A similar expression applies to $(t + 1/t)$ with τ replaced by $-\tau$. We have shown that

$$f_1 = \frac{(f_e\tau+1)}{\tau+f_e}, \tag{68.1}$$

$$f_2 = \frac{(f_e\tau-1)}{\tau-f_e}. \tag{68.2}$$

If we define $c_1 = \cos(A_1)$ and $c_2 = \cos(A_2)$, then Eq. 62 yield

$$\cos(B_1) = \frac{c_1\tau+1}{\tau+c_1}, \tag{69.1}$$

$$\cos(B_2) = \frac{c_2\tau-1}{\tau-c_2}. \tag{69.2}$$

Hence, $\zeta_1 = \zeta_2$ when

$$(f_e\tau + 1)(\tau - f_e)(c_1\tau + 1)(\tau - c_2) = (f_e\tau - 1)(\tau + f_e)(c_2\tau - 1)(\tau + c_1). \quad (70)$$

We define

$$\phi \equiv \frac{\frac{c_1+c_2}{2}(f_e - \frac{1}{f_e}) - (1 - c_1c_2)}{c_1 - c_2}. \quad (71)$$

Then Eq. 70 reduces to

$$\tau^4 - 2\phi(\tau^3 - \tau) - 1 = 0 \quad (72)$$

or

$$(\tau^2 - 1)(\tau^2 - 2\phi\tau + 1) = 0. \quad (73)$$

We seek a value of β which is less than e. By Eq. 67, τ must be greater than unity. Thus, the roots $\tau = \pm 1$ are extraneous. We next demonstrate that $\phi > 1$: Let $c_1 = (1 + r)c_2$, $r > 0$, and $f_e c_2 = 1 + p$, $p > 0$. Then

$$\phi = \frac{(1 + \frac{r}{2})(1 + p - \frac{c_2^2}{1+p}) - 1 + (1 + r)c_2^2}{rc_2}$$

$$> \frac{p + \frac{r}{2} + \frac{r}{2}c_2^2}{rc_2} > \frac{1}{2}\left(c_2 + \frac{1}{c_2}\right) > 1.$$

As p approaches zero, ϕ approaches $\frac{1}{2}(c_2 + \frac{1}{c_2})$. The only root of Eq. 73 greater than unity is thus

$$\tau = \phi + \sqrt{\phi^2 - 1}. \quad (74)$$

From Eq. 67, we obtain

$$\beta = (\tau + \sqrt{\tau^2 - 1})^{-1}. \quad (75)$$

By way of illustration, consider $S_1(0.1, 10, 0°)$ and $S_2(0.1, 10, \cos^{-1}\frac{1}{f_{0.1}})$, where $f_{0.1} = (0.1 + 10)/2 = 5.05$. The optimum single parameter is $w = 1$ with associated error reduction of $R_1 = R_2 = \frac{1-0.1}{1+0.1} = \frac{0.9}{1.1}$. Hence, $R = R_1R_2 = \frac{0.81}{1.21}$ is the error reduction when $J = 1$. The transformation with $\beta = 0.1$ yields $s = \frac{0.1+0.1}{1+0.01} = \frac{0.2}{1.01}$ with $S_1(s, 1/s, 0°)$. The error reduction over the infinite disc is $R_2 = 1$ and the error reduction over the transformed region 1 is $R_1 = \frac{1-s}{1+s} = \frac{0.81}{1.21}$. The back transformation of $w = 1$ remains at $w = 1$ and it is no surprise that this optimum single parameter is invariant. Although R_1 and R_2 change, the product is invariant when we back transform.

When $J = 1$ there is no need for further alignment. For $J > 1$, however, choice of optimum parameters is facilitated by the transformation. Consider the case of $J = 2$. Repeated use of $w = 1$ squares the reduction to $R = 0.448$. The optimum two parameters over region S_1 yield $R_1 = 0.384$ (determined as described in Chap. 1) while R_2 is greater than the value obtained for the disc with its optimum parameters of $w_1 = w_2 = 1$ which is 0.669 so that $R > 0.669 \times 0.384 = 0.257$, with a possible improvement over $w_1 = w_2 = 1$. Improvement is guaranteed by transforming to determine the truly optimum elliptic-function parameters. The transformed interval for S_1 is (0.198,5.05) for which the optimum two parameters yield $R_1 = 0.2366$. Although $R_2 = 1$, the product is now $R = 0.2366$. As the cycle length J increases, greater improvement is achieved through use of the optimum cycle in the transformed space.

In general, the product $R = R_1 R_2$ remains fixed after the back transformation even though the individual values change. Error reduction in the presence of complex spectra is analyzed in Sect. 4.5. Two parameters play a crucial role. One is the nome, q, of the elliptic-function region. This is a function of ζ only and is the same for both spectra after alignment. The other, $\nu \in (0, 1]$, depends on both the real interval and the angle. The number of iterations needed to yield a prescribed error reduction varies inversely as ν, which is a measure of the retardation in convergence as the angle increases. An approximate value is $\nu \simeq (1 - A^\circ/90^\circ)$, and the precise value is computed as described on pp.77–78. Asymptotic error reduction per iteration (as J increases) varies as

$$R^{1/J} \sim q^{(\nu_1+\nu_2)}. \tag{76}$$

A prescribed error reduction ε is obtained by choosing

$$J > \frac{\ln \frac{\varepsilon}{4}}{(\nu_1 + \nu_2)\ln q}. \tag{77}$$

For this J, a more accurate estimate of the error reduction is given by $R = R_1 R_2$ where

$$R_k = q^{\nu_k J} \frac{1 + q^{2J(1-\nu_k)}}{1 + q^{2J(1+\nu_k)}}. \tag{78}$$

When the spectra can be embedded in elliptic-function regions the parameters determined by this method are close to optimum. Comparison with parameters determined by methods described by [Istace] and [Starke] should be of great interest.

4.6.5 Alignment When One Spectrum Is a Disk

We now address the case where either value is equal to unity. The corresponding spectrum is a disk. It is known that the transformations in Eq. 58 retain a disk spectrum. We will prove this while establishing properties useful in alignment.

Theorem 15. *The transformation in Eq. 58 transforms a disk, which is characterized by $\zeta = 1$, into a disk. Thus, $\zeta = 1$ is invariant.*

Proof. The theorem may be established with either the transformation or its inverse. We choose $s > s_0$ so that $\beta > 0$, and replacing $\cos\theta$ by ζ/f in Eq. 62, we have

$$\cos\theta_s = \frac{f + \frac{\zeta}{2}(\beta + 1/\beta)}{\zeta + \frac{f}{2}(\beta + 1/\beta)}, \tag{79}$$

where $f = f_{s0}$ and $\zeta = \zeta_{s0}$. Hence,

$$\zeta_s = f_s \cos\theta_s = f_s \frac{f + \frac{\zeta}{2}(\beta + 1/\beta)}{\zeta + \frac{f}{2}(\beta + 1/\beta)}.$$

One can apply Eq. 61 to establish the identity

$$\frac{1}{2}(\beta + 1/\beta) = \frac{f_s f - 1}{f - f_s}.$$

It follows that

$$\zeta_s = f_s \left[\frac{f + \zeta\left(\frac{f_s f - 1}{f - f_s}\right)}{\zeta + f\left(\frac{f_s f - 1}{f - f_s}\right)} \right], \tag{80}$$

and when $\zeta = 1$, $\zeta_s = f_s \frac{f^2 - 1}{f_s f^2 - f_s} = 1$. □

When either spectrum is a disk, the two-variable problem is reduced to one variable by a simple but elegant method. As s approaches $1/p$, the radius of the disk increases. In the limit, when we choose optimal parameters for $S_1(1/p)$, the error function has absolute value unity on the boundary of the infinite disk. Hence, the back transformation will have constant absolute value on the boundary of S_2 and theory establishes this as sufficient for simultaneous parameter optimization for the two spectra. When $s = 1/p$, $\beta = t_0$ and angle θ_s is determined with Eq. 62.

It is instructive to consider the alignment equations when $p = 0$. From Eq. 73, $\phi = (\tau + \frac{1}{\tau})/2$, $\tau > 1$. Hence, $\tau = \frac{1}{c_2} = f_e$. From Eq. 69.2, $B_2 = 90°$. Also, from Eq. 75, $\beta = (f_e + \sqrt{f_e^2 - 1})^{-1} = (\frac{e+1/e}{2} + \sqrt{(\frac{1/e-e}{2})^2})^{-1} = e$.

When both spectra are disks, the transformation leaves $S_1(1/p)$ as a disk with optimal iteration parameters all equal to unity. The back transformation yields optimal values for **p** and **q** for the two disks.

Careless application of the theory can lead to erroneous conclusions. One pitfall will now be illustrated by example. Let S_1 be the line $\cos\theta + i\sin\phi$, $|\phi| \le \theta$. Let S_2 be a disk with real intercept $[\cos\alpha, \sec\alpha]$. The optimum single parameter for both spectra is unity. The corresponding error reduction is

$$R = \left(\frac{1 - \cos\alpha}{1 + \cos\alpha}\right) \tan\frac{\theta}{2}.$$

Note that as θ approaches zero the line shrinks to a point and R approaches zero. We may subtract $\cos\theta$ from S_1 and add $\cos\theta$ to S_2. Then for any real parameter the absolute value of the error reduction is unity along the shifted line which now falls on the imaginary axis. If we choose as the parameter the square root of the endpoints of the real intercept of the translated disk, then the error reduction has a constant absolute value on the boundary of the disk. As θ approaches zero, the translation approaches unity and the error reduction over the disk approaches

$$R = \left(\frac{1-\sqrt{\cos\alpha}}{1+\sqrt{\cos\alpha}}\right).$$

This is not zero and certainly not optimal for these spectra. Yet the error reduction has constant absolute value along both boundaries. The flaw in the argument is that the roots of the error function do not lie within spectrum S_1. Rouché's theorem cannot be applied. The line in this example is not an elliptic-function region. The analysis in this section applies only to spectra embedded in elliptic-function regions.

When θ approaches $\pi/2$, error reduction along the line is slight and the shift of S_1 to the imaginary axis yields nearly optimal parameters. This works for any S_2 which may be embedded in an elliptic-function region after the shift. In general, if one region dominates the other in error reduction, one may shift the weaker region until the smallest real component of its shifted eigenvalues is zero and compute parameters which are optimal for the shifted dominant region.

When $(\zeta_1 - 1)(\zeta_2 - 1) < 0$, the spectra of $S_1(s)$ and $S_2(t)$ are always separated by a disk and cannot be aligned with Eq. 58. The simplest resolution is to not align and just use parameters for the spectrum with $\zeta < 1$.

4.6.6 Illustrative Examples

We now illustrate some of the algorithms with a few simple examples. The real intervals $S_1 = [0.1, 10]$ and $S_2 = [10, 100]$ transform as in Part 1 into the interval $[0.19967, 1]$ for which the optimum single parameter is $w_1 = 0.44684$. If $\theta_1 = 78.58°$ and $\theta_2 = 54.9°$, the complex regions are disks and repeated use of parameters back transformed from w_1 is optimal. We compute these values as $q_1 = 4.1$ and $p_1 = 20.74$. The error reduction per iteration assumes its largest magnitude at the interval endpoints and is $\varepsilon = 0.1462$. If one were to arbitrarily try the geometric means of the intervals ($q_1 = 1$ and $p_1 = 10^{3/2}$) as parameters, one would find that the error at the point $z = 10$ is equal to $\varepsilon = 0.4287$. This may not even be the bound, but it is certainly not nearly as small as the optimal value.

We now illustrate our algorithms for the case where one of the spectra is a disk. We obtain optimum parameters for $J = 2$ by transforming the disk to infinite radius. Let $S_1 = (1/4, 4, 0°)$ and $S_2 = (1/2, 2, 36.87°)$. We compute $\zeta_2 = 1$, thereby establishing that S_2 is a disk. By Eq. 59, $p = 1.5$, and we realign at $s = 1/p = 2/3$. The transformation from $s_0 = 0.25$ to $s = 2/3$ is attained with $\beta = 1/2$.

(The inverse is with $\beta = -1/2$.) The best parameter for $J = 1$ is unity, which transforms back into $p_1 = q_1 = 1$ with corresponding error reduction of $R = 0.2$. This was known from the given spectra, and required no transformation. The best two parameters for $S_1(2/3) = (2/3, 3/2, 0°)$ are $w_1 = 0.7522$ and its reciprocal $w_2 = 1.3295$. We compute $q_1 = (w_1 - 0.5)/(1 - 0.5w_1) = 0.4042$ and $q_2 = 1/q_1 = 2.474$. We compute $p_1 = (w_1 + 0.5)/(1 + 0.5w_1) = 0.90995$ and $p_2 = 1/p_1 = 1.099$. We compute the error reduction at various points on the boundaries of S_1 and S_2 and ascertain that the reduction is indeed a constant value of $R = 0.02$.

Chapter 5
Lyapunov and Sylvester Matrix Equations

Abstract ADI iterative solution of Lyapunov and Sylvester matrix equations may be enhanced by availability of a stable algorithm for similarity reduction of a full nonsymmetric real matrix to low bandwidth Hessenberg form. An efficient and seemingly stable method described here has been applied successfully to an assortment of test problems. Significant reduction in computation for low rank right-hand sides is possible with ADI but not with alternatives. Initial analysis by Penzl was improved upon by Li and White. A Lanczos algorithm is exposed here for approximating a full matrix by a sum of low rank matrices. This is especially useful in a parallel environment where the low rank component solutions may be computed in parallel.

5.1 Similarity Reduction of Nonsymmetric Real Matrices to Banded Hessenberg Form

In Chap. 3 Sect. 3.8 it was observed that the search for an efficient and robust similarity reduction of a real matrix to banded Hessenberg form was motivated by application to ADI iterative solution of Lyapunov and Sylvester matrix equations. One promising candidate will now be described. The limited numerical studies thus far performed have been encouraging.

Any real $n \times n$ symmetric matrix may be reduced to a similar symmetric tridiagonal matrix in $2n^3/3$ flops by successive Householder (HH) transformations [Golub and vanLoan, 1983]. All eigenvalues may then be computed with another $2n^3/3$ flops. Any unsymmetric real matrix may be reduced similarly to upper Hessenberg form with $5n^3/3$ flops [Golub and vanLoan, 1983 (p. 223)]. All eigenvalues may now be computed with the implicit QR algorithm in around $8n^3$ flops [Golub and vanLoan, 1983 (p. 235)]. Half as many flops are required when gaussian rather than HH transformations are used. However, greater stability of HH has led to its use. The MATLAB program EIG computes eigenvalues in this manner.

E. Wachspress, *The ADI Model Problem*, DOI 10.1007/978-1-4614-5122-8_5,
© Springer Science+Business Media New York 2013

A banded matrix may be stored in sparse form. Eigenvalues of a sparse matrix may be computed with the MATLAB EIGS program. This program uses an Arnoldi iteration and all eigenvalues may be found with $O(bn^2)$ flops, where b is the upper bandwidth. This possibility stimulated search for stable similarity reduction of unsymmetric matrices to banded form. Early studies [Dax and Kaniel, 1981] indicated that eigenvalues of a tridiagonal matrix obtained by gaussian reduction of an unsymmetric matrix with unbounded multipliers were fairly accurate. Although this eliminated the $8n^3$ flops of the QR algorithm, loss of stability, accuracy, and robustness resulted in few applications. More recent research was directed toward gaussian reduction to banded upper Hessenberg form with limited gaussian multipliers [Geist, Lu and Wachspress, 1989; Howell and Geist, 1995]. The BHESS program [Howell, Geist and Diaa, 2005] culminating these studies was a promising alternative to QR. However interaction of large gaussian multipliers limited accuracy and stability. Relationship between bandwidth, multiplier bound, and accuracy was not amenable to analysis. QR reduction to upper Hessenberg form followed by the implicit QR algorithm with a double Wilkinson shift as in the MATLAB EIG program remained the method of choice. In 1995 I suggested a possible improvement of BHESS which seems not to have been programmed until my QGBAND in 2011. When applied to symmetric matrices this algorithm is identical to the standard HH reduction. It reduces a nonsymmetric real matrix to a banded upper Hessenberg matrix. The reduction progresses from row $k = 1$ to $n-2$. In all numerical studies reported here an input matrix A was first normalized to $S_0 = \text{snorm} * A$ where snorm $= 1/\sqrt{\|A\|_1 \|A\|_\infty}$. This results in a bound of unity on the spectral radius of S_0. The 1-norm of S_0 was close to unity for all random matrices reported here. Let matrix S_0 reduced through $k-1$ be S_{k-1} and let $S \equiv S_{n-2}$. Let $k(1)$ be the first row in S_{k-1} with a nonzero element beyond col $k\,V$:

```
D X 0...............................0
X D X 0 ..........................0
0 X D X X 0....................0
0 0 X D X X 0.................0
0 0 0 X D X X 0..............0
0 0 0 0 X D X X..............X        k(1)
0 0 0 0 0 X D X...............X        k
0 0 0 0 0 0 X D X............X
0 0 0 0 0 0 X X D X.........X
.......................................
0 0 0 0 0 0 X X.................D
```

Column k is now reduced to zero below row $k+1$ with a HH step:

```
D X 0...............................0
X D X 0 ...........................0
0 X D X X 0.....................0
0 0 X D X X 0..................0
0 0 0 X D X X 0..............0
0 0 0 0 X D X..................X        k(1)
0 0 0 0 0 X D X................X        k
0 0 0 0 0 0 X D X.............X
0 0 0 0 0 0 0 X D X..........X
.......................................
0 0 0 0 0 0 0 X.................D
```

For the standard HH reduction to upper Hessenberg form this yields S_k and column $k+1$ is then reduced. Further row treatment is generally required for banded reduction. All rows from $k(1)$ on may now have nonzero entries beyond column k. A HH reduction to reduce row $k(1)$ to zero beyond column $k+1$ would reintroduce nonzero elements below row $k+1$ in column k. The HH row reduction is chosen instead to reduce row $k(1)$ beyond column $k+2$:

```
D X 0...............................0
X D X 0 ...........................0
0 X D X X 0.....................0
0 0 X D X X 0..................0
0 0 0 X D X X 0 0...,,,,......0
0 0 0 0 X D X X X 0........0        k(1)
0 0 0 0 0 X D X...............X        k
0 0 0 0 0 0 X D X............X
0 0 0 0 0 0 0 X D X..........X
.......................................
0 0 0 0 0 0 0 X.................D
```

This leaves column k unchanged. The only nonzero entries in row $k(1)$ beyond column k are $S_{k(1),k+1}$ and $S_{k(1),k+2}$. A bound M is specified on the magnitude of gaussian multipliers. If the magnitude $|S_{k(1),k+2}/S_{k(1),k+1}|$ is greater than M, there is no further row reduction before proceeding to reduction of column $k+1$. The bandwidth is increased by one. On the other hand, if the ratio is less than M element $S_{k(1),k+2}$ is reduced to zero with a gaussian step:

```
D X 0.................................0
X D X 0 ...........................0
0 X D X X 0......................0
0 0 X D X X 0..................0
0 0 0 X D X X 0...............0
0 0 0 0 X D X X 0............0          k(1)
0 0 0 0 0 X D X................X          k
0 0 0 0 0 0 X D X.............X
0 0 0 0 0 0 0 X D X..........X
...........................................
0 0 0 0 0 0 0 X.........x.......D
```

After this step, $k(1)$ is increased to $k(1)+1$ and the algorithm progresses to $k+1$. (In this example, the fact that $k(1)$ was $k-1$ meant that the bandwidth was increased before step k from one to two nonzero elements beyond the diagonal. Each time the width is increased $k-k(1)$ increases by one. Thus, the two rather than one elements to the right of the diagonal at $k(1)$ after reduction at k were not an increase at k.) The QG reduction requires $8n^3/3$ flops. However, eigenvalues of the banded matrix can be computed in $O(bn^2)$ flops instead of the $8n^3$ flops of QR.

In earlier attempts the entire reduction was done with gaussian transformations. Later attempts used HH reduction of columns followed by gaussian reduction of rows but the preliminary HH row reduction was not performed. There is significantly less interaction of gaussian multipliers with this preliminary HH step to reduce row $k(1)$ to zero beyond element $(k(1),k+2)$. A measure of the stability of the reduction from S_0 to S is the ratio $||S||_1/||S(0)||_1$.

Results of two numerical studies illustrate characteristics of the QG reduction Table 5.1. Eigenvalues of S differed from those of S_0 by $O(10^{-11})$ in all cases. The first matrix was a random matrix with $n = 100$ and the second was a random matrix of order 300. Extra is the number of nonzero entries beyond the tridiagonal:

Table 5.1 Effect of multiplier bounds

n	M	Extra	$\|S\|$
100	10,000	0	178
100	1,000	0	178
100	100	171	45
100	10	2,551	50
300	10,000	239	1,557
300	1,000	317	87
300	100	4,870	80
300	10	36,550	53

For $n = 100$ a bound of $M = 1,000$ sufficed to reduce to tridiagonal form. For $n = 300$ the bound of $M = 1,000$ led to extra nonzero elements beyond the tridiagonal but a relatively small increase in the norm. Although $M = 10,000$

led to fewer elements beyond the tridiagonal the norm increased significantly. The eigenvalues remained accurate. The default bound of $M = 1,000$ was chosen on the basis of this and other numerical studies.

Studies were performed on a PC and matrices of large order required significant memory and computing time. A random matrix of order 1,000 was reduced in about ten minutes. The default value of $M = 1000$ resulted in 15 band increases with a total of 8,941 nonzero elements beyond the tridiagonal. The 1-norm of S was 113.5. The absolute values of the eigenvalues ranged from 2×10^{-4} to 0.037 and the reduced matrix values differed from the true values by less than 5×10^{-10}. These preliminary studies suggest that QG is a robust algorithm for reducing a full matrix to low-bandwidth upper Hessenberg form with accurate retention of eigenvalues. More extensive numerical studies are needed.

5.2 Application of QG to the Lyapunov Matrix Equation

My recognition in 1982 that the Lyapunov matrix equation is a model ADI problem initiated extensive development of methods for solving Lyapunov and Sylvester equations. [Benner, Li and Truhar, 2009] provided an excellent overview of this effort. They introduced a projection method as a competitive alternative to other methods for a wide class of problems of practical concern. They also observed that the existence of an algorithm for efficient reduction of a full matrix to banded Hessenberg form as a prelude to ADI iterative solution could be quite efficient. The QG algorithm seems to offer that possibility. Direct solution by ADI requires solution of two linear systems of order n for each iteration. This requires around $4n^3/3$ flops so that J iterations require around $4Jn^3/3$ flops. The break-even point with B–S is when $4J/3 = 20$ or $J = 15$. Efficient ADI iteration requires knowledge of the spectrum of A which may itself be computation intensive. Once the ADI linear system for a two-step iteration has been factored in $2n^3/3$ flops, the back-substitution stage on the n columns of the right-hand side can be performed in parallel. This leads to $2Jn^3/3$ flops or a break-even number of $J = 30$. The number of ADI iterations required to achieve prescribed accuracy varies as the log of the condition of matrix A.

The HH similarity G that reduces the Lyapunov matrix A to upper Hessenberg form S requires $5n^3/3$ flops. The QG transformation from A to the banded S with two HH transformations at each step requires $8n^3/3$ flops. One may either store G or the HH and gauss transformations in n^2 words of memory. Transforming the right-hand side C to H requires another $10n^3/3$ flops for a total of $6n^3$ flops to reduce Eqs. 1.1 and 1.2 of Chap. 3 to Eq. 2 of Chap. 3. This compares with $25n^3/6$ flops for the B–S transformation. Recovery of X from Y requires another $20n^3/9$ flops when one takes advantage of symmetry. A total of $74n^3/9$ flops are needed. B–S recovery requires $10n^3/9$ flops for a total of $55n^3/9$ flops. The B–S solution of the transformed equation is also $O(n^3)$ and is a major part of the computation, leading to a total of around $20n^3$ flops.

The only eigenvalues needed for determining effective ADI iteration parameters are those with small real part and one of largest magnitude (the spectral radius of S). After QG reduction to banded form the sparse matrix MATLAB EIGS program may be used to compute these crucial values efficiently. When S is of order n one may compute $O(n^{1/2})$ small eigenvalues and $O(n^{1/4})$ values of largest magnitude. This is an $O(n^{3/2})$ computation. Solution of Eq. 2 of Chap. 3 for Y is $O(bn^2)$. Thus, even though the transformation to banded form with QG requires around 50% more flops than transformation to Hessenberg form, the overall $O(n^3)$ flop count for ADI–QG is around 41% of the B–S value.

In a parallel environment ADI iteration has another advantage. Although the transformation can be performed in parallel for both reduction algorithms, columns in each of the two ADI iteration steps can be computed in parallel while the B–S solution of the transformed equations is less amenable to parallel computation.

5.3 Overview of Low-Rank Right-Hand Sides

The ADI method can take advantage of low-rank right-hand side C in the Lyapunov equation. This of not possible with the alternative methods thus far used. Low-rank right-hand sides recur in application. Let R be an $n \times m$ matrix with $m << n$. Then $C = RR'$ in Eq. 1 is of rank m. Penzl [Penzl, 1999] observed that low-rank Lyapunov equations could be solved more efficiently with ADI iteration. Similar savings could not be realized with the B–S algorithm. Subsequently, Penzl's algorithm was improved in [Li and White, 2002]. The Li–White (LW) algorithm requires solution of one linear system of order n with an $n \times m$ right-hand side for each ADI iteration. Li and White did not transform from Eqs. 1.1 and 1.2. They considered sparse A and approximated solution of the ADI linear systems by an iteration with programs like GMRES. In a parallel environment the number of processors to solve for all columns in parallel is decreased from n to m. The B–S algorithm cannot take full advantage of low-rank right-hand sides. The LW approach without preliminary transformation to banded form has been adopted by some practitioners. The Sylvester matrix equation was discussed in Eqs. 9.1 and 9.2 of Chap. 3. Low-rank equations may be treated in similar fashion to low-rank Lyapunov equations [Wachspress, 2008].

5.4 The Penzl Algorithm

Penzl's algorithm will now be described. The low-rank Lyapunov equation $AX + XA^\top = CC^\top$ may be reduced to

$$SY + YS^\top = H, \tag{1}$$

where $S = GAG^{-1}$, $Y = GXG^{\top}$, and $H = FF^{\top}$ with $F = GC$ of rank $r << n$. Matrix G^{-1}, used to recover X from Y, is accumulated during the QG reduction of A to S. The nonfactored ADI iteration equations with $Y_o = 0$ and the number of iterations J determined from the spectrum and a prescribed bound on the solution error are

$$[S + w_j I]Y_{j-1/2} = H + Y_{j-1}[w_j I - S]^{\top}, \tag{2.1}$$

$$[S + w_j I]Y_j = H + Y_{j-1/2}^{\top}[w_j I - S]^{\top}, \tag{2.2}$$

for $j = 1, 2, \ldots, J$. For each value of j, matrix $S + w_j I$ is factored, and the $2n$ linear systems for the columns of $Y_{j-1/2}$ and then Y_j are solved. The number of iterations J is often $O(\log n)$. The reduction to banded form and recovery of X from Y are the $O(n^3)$ steps. Although Y_j is symmetric, $Y_{j-1/2}$ is in general not symmetric. If the iteration matrix for the j-th step is defined as

$$R_j = [w_j I + S]^{-1}[S - w_j I], \tag{3}$$

then

$$Y_j = 2w_j[w_j I + S]^{-1}H[w_j I + S]^{-\top} + R_j Y_{j-1} R_j^{\top}. \tag{4}$$

Equation 4 provides the basis for improved efficiency when $r << n$. Let the ADI approximation after iteration j with parameter w_j be Y_j. Suppose

$$Y_j = \sum_{i=1}^{j} Z_i(j) Z_i(j)^{\top} \tag{5}$$

Then, with $Z_i(0) = 0$,

$$Z_j(j) = \sqrt{2w_j}[w_j I + S]^{-1}F \tag{6.1}$$

$$Z_i(j) = R_j Z_i(j-1) \qquad i = 1, 2, \ldots, j-1. \tag{6.2}$$

Now the ADI iteration of Eqs. 6.1 and 6.2 replaces Eqs. 2.1 and 2.2. We note that Y_j is of rank jr. Note that $Z_i(j-1)$ is a matrix of order nrj, so the algorithm requires solving $rJ(rJ+1)/2$ linear systems of order n.

The reduced equation form when complex parameters are used is maintained by rewriting Eq. 2 as

$$[S + w_j I]Y_{j-1/2} = H + Y_{j-1}[w_j I - S^{\top}], \tag{7.1}$$

$$Y_j[S^{\top} + w_j' I] = H + [w_j' I - S]Y_{j-1/2}, \tag{7.2}$$

where w' is the complex conjugate of w. Then for this iteration, Eqs. 3 and 6 become

$$Q_j = [w_j I + S]^{-1}[S - w_j' I], \tag{8.1}$$

$$Z_j(j) = \sqrt{w_j + w_j'}[w_j I + S]^{-1}F, \tag{8.2}$$

$$Z_i(j) = Q_j Z_i(j-1) \qquad i = 1, 2, \ldots, j-1. \tag{8.3}$$

We observe that when $w_{j+1} = w'_j$, $Q_j Q_{j+1} = R_j R_{j+1}$. Thus, the ADI iteration matrix is recovered when the roles of w and w' are interchanged for the iteration with the conjugate parameter. Since this introduces complex Z_j and complex arithmetic requires more flops than real arithmetic, the complex iteration parameters are saved for last. Repeated use of a single real parameter with more iterations may sometimes be more efficient than the "optimal" set of complex parameters. In some cases, it is best to apply complex parameters only to reduce error associated with eigenvalues close to the imaginary axis and to embed the remaining spectrum in a disk for which repeated use of a real parameter is optimal.

A measure of solution accuracy is found by plugging it into the equation. For the full system one computes AY and the residual error $||H - AY - YA^{\top}||_1$. For the low-rank system, $W_i = SY_i$ is first computed for each i and then $U_i = W_i Y_i^*$ is summed to yield SY.

5.5 The Li-White Algorithm

The Li-White recursive algorithm will now be described. Reduction to banded upper Hessenberg form was common to all Lyapunov solvers studied. By Eq. 8.2,

$$Z_J(J) = \sqrt{w_J + w'_J}[w_J I + S]^{-1} F, \tag{9.1}$$

$$Z_{J-1}(J-1) = \sqrt{w_{J-1} + w'_{J-1}}[w_{J-1} I + S]^{-1} F. \tag{9.2}$$

By Eqs. 9.1 and 9.2,

$$\begin{aligned} Z_{J-1}(J) &= [w_J I + S]^{-1}[S - w'_J I] Z_{J-1}(J-1) \\ &= \sqrt{\frac{w_{J-1} + w'_{J-1}}{w_J + w'_J}}[w_{J-1} I + S]^{-1}[S - w'_J I] Z_J(J). \end{aligned} \tag{10}$$

In general, proceeding back from $i = J$ to 1, the r columns of each $Z_i(J)$ are computed in succession:

$$Z_J(J) = \sqrt{w_J + w'_J}[w_J I + S]^{-1} F, \tag{11.1}$$

$$Z_{i-1}(J) = \sqrt{\frac{w_{i-1} + w'_{i-1}}{w_i + w'_i}}[I - (w'_i + w_{i-1})(S + w_{i-1} I)^{-1}] Z_i(J), \tag{11.2}$$

$$i = J, J-1, \ldots, 2.$$

The result is independent of parameter ordering. Complex arithmetic is reduced by ordering all complex parameters ahead of real parameters since the algorithm starts with w_J and proceeds backward. Now each iteration requires solution of only r linear systems for a total of rJ systems rather than the $rJ(rJ + 1)/2$ required of the Penzl algorithm.

5.6 Sylvester Equations

This approach also applies to Sylvester equations. Consider the banded system of order $n \times m$:

$$SY + YT = EF^{\top} \tag{12.1}$$

of the Sylvester equation

$$AX + XB = H, \tag{12.2}$$

where $H = CD^{\top}$, E is order $n \times r$, and F is order $m \times r$. The transformation matrices L_s and L_t are saved for computing $X = L_s Y L_t$. The ADI approximation to the solution after j iterations is

$$Y_j = \sum_{i=1}^{j} U_i(j) V_i(j). \tag{13}$$

Now matrices $U_i(j)$ and $V_i(j)$ must be computed for each j. There are in general two iteration parameters, u_j and v_j, for each iteration j. Matrix R_j of Eq. 3 is now

$$R_j = [u_j I + S]^{-1}[v_j I - S], \tag{14}$$

and similarly

$$Q_j = [v_j I + T]^{-1}[u_j I - T]. \tag{15}$$

Now Eq. 4 become

$$[S + u_j I]Y_{j-1/2} = EF^{\top} + Y_{j-1}[u_j I - T], \tag{16.1}$$

$$Y_j[T + v_j I] = EF^{\top} + [v_j I - S]Y_{j-1/2}. \tag{16.2}$$

The recursion formulas for the $U_i(j)$ are

$$U_i(j) = R_j U_i(j-1), \quad i = 1, 2, \ldots, j-1, \tag{17.1}$$

$$U_j(j) = (u_j + v_j)[u_j I + S]^{-1} E, \tag{17.2}$$

and

$$V_i(j) = Q_j^* V_i(j-1), \quad i = 1, 2, \ldots, j-1, \tag{18.1}$$

$$V_j(j) = [v_j^* I + T']^{-1} F. \tag{18.2}$$

The Li-White algorithm may be introduced to reduce iteration complexity.

5.7 Approximating a Full Matrix by a Sum of Low-Rank Matrices

The analysis applies when the right-hand side of the reduced Lyapunov equation (Eq. 1) is of the form $H = FPF^{\top}$ with P of order $r \times r$ or the r.h.s. of the reduced Sylvester equation (Eq. 12.1) is $EPF^{\top}$. The matrix P does not affect the ADI iteration equations. Significant reduction in computation may be realized if the r.h.s. can be approximated reasonably well in this form. A review of matrix factorization with discussion of low-rank approximation is given in [Hubert et al., 2000]. Penzl observed that "splitting up the right hand side matrix into a sum of low-rank matrices enables an efficient parallelization of [his] method."

This suggests a general procedure for solving all N-stable Lyapunov problems. Let the $n \times m$ matrix of orthonormal vectors of m Lanczos steps applied to matrix H be K and let the Lanczos coefficients determine the tridiagonal matrix T of order m. Then the matrix $F_1 = KTK^{\top}$ is a rank m approximation to H. We may, therefore, compute $H_1 = H - F_1$ and perform m Lanczos steps on H_1 with initial vector equal to what would have been the $m+1$ vector of the previous Lanczos steps on H. This may be continued to yield a set of F_j. The norms of successive H_j should decrease and the algorithm may be terminated when sufficient accuracy is achieved with the sum of the low-rank approximations. If H is of full rank with $n = 100$ and m is 5, for example, the algorithm should terminate when j is around 20. The matrix H_{21} will in general not be the zero matrix since the Lanczos vectors from each H_j are not orthogonal to the previous vectors. When the sum of the low-rank subspaces approaches H, the rank of subsequent subspaces may be smaller than m.

In a pilot MATLAB program, when the absolute value of the $(k, k+1)$ element of T (for $k < m$) was less than 0.001 times that of element (1,2), the rank of the subspace was chosen as k. The algorithm was terminated when the order of T was one. Having generated a set of low-rank matrices, one may solve the low-rank Lyapunov systems in parallel. In this application, the matrix A and its transformation to S is common to all low-rank problems. The reduction, spectrum evaluation, ADI iteration parameter determination, and back transformation need only be done once. This approach extends application of the Li–White algorithm to general right-hand sides. When H is not given in factored form but is of rank $r << n$ the Lanczos partitioning will expose the rank deficiency of H and may lead to more efficient solution.

A set of test problems was considered with the matrix B in Eq. 12.1 chosen as $A + A'$. This B is symmetric but not necessarily SPD. In general, the unique solution X need not be SPD when B is not SPD. However, X is the identity matrix for this choice of B. It should be noted that the tridiagonal matrices T need not be SPD since only K appears as a rhs for the ADI iterations.

One test case was run with a random N-stable A of order 30 for which B (and hence the rhs H of the transformed equation) was not SPD. The initial value for m was chosen as 5. The first seven subspaces were of rank 5, but the eighth was of rank 1. The sum of the low-rank solutions agreed with the true solution to four significant places. Another problem was solved with A of order 100 and $m = 10$. The first 11 subspaces were of rank 10 and the 12th was of rank 1. The factored Y agreed with the nonfactored Y to four significant figures. Comparable accuracy was obtained with 14 subspaces of rank $m = 8$ and a 15th of rank 1. For this problem, an initial choice of $m = 12$ resulted in nine subspaces of rank 12, a tenth of rank 5, and a last subspace of rank 1. The factored result agreed with the nonfactored result to five significant places. To illustrate how the Lanczos algorithm exposes rank deficiency of a given full matrix, a random full matrix of order 100 and rank 20 was chosen as B. The value for m was chosen as 5. The algorithm terminated with four subspaces of rank 5, one of rank 4, and the last of rank 1.

An in-depth comparison of computation time for various approaches should be made. Many stages are easily parallelized. Reduction to banded upper Hessenberg form, the Lanczos algorithm, GMRES-type solution of the ADI iteration equations with sparse A, and simultaneous solution for all low-rank matrices generated by the Lanczos algorithm applied to a full-rank system are among these stages. It should be noted that the ADI iteration equations may be solved for all columns of X or Y simultaneously once the right-hand sides of the equations are computed. However, computation of these right-hand sides for a full rank H of order n requires a factor of n/m times computation with the $n \times m$ factor K in the Li–White algorithm. When the coefficient matrix A is sparse, solution without reduction may be best, provided one can determine good ADI iteration parameters. However, when A is not sparse, the GMRES approach may become less efficient even with good ADI parameters. Optimization and comparison of relative efficiency of the various methods is a fertile area for further research.

5.8 Summary

The ADI iteration originally proposed by Peaceman and Rachford in 1955 for numerical solution of difference equations for elliptic partial differential equations has been widely used for solving such problems. Analysis is less precise when the coefficient matrix is split into noncommuting components for the iteration. Almost thirty years after inception of ADI iteration, it was recognized that Lyapunov and Sylvester matrix equations have the commutation property. This led to generalization of the ADI iteration theory from real to complex spectra with a rather elegant application of the theory of modular transformations of elliptic functions. It also stimulated research into similarity reduction of a full nonsymmetric real matrix to sparse form and in particular to banded upper Hessenberg form. Despite the rapid convergence of ADI iteration for these problems, earlier methods of Bartels–Stewart and Smith remained competitive and were already incorporated in major software packages. After another twenty years had elapsed, it was observed by Penzl that low-rank equations could be treated by a low-rank ADI iteration more efficiently than by conventional methods which have not been shown to admit significant gains in efficiency for low-rank problems. The subsequent contribution by Li and White further reduced computational effort so that ADI iteration is clearly superior to the other methods for such problems. Relative merits of iterating with the full system and iterating after reduction to banded Hessenberg form are still under consideration and may well depend on specific applications.

The possibility of approximating a general right-hand side by a sum of low-rank matrices extends application of the Li–White algorithm to all N-stable Lyapunov systems. The Lanczos algorithm has worked well for generating a sum of low-rank approximations in the few test problems thus far considered. In general, ADI solution of Lyapunov equations by any of the methods discussed is well suited for parallel computation.

Chapter 6
MATLAB Implementation

Abstract Theory in the previous chapters was implemented with a set of MATLAB programs which are described here. These programs were verified with a small set of test problems.

6.1 MATLAB to Expedite Numerical Studies

[1]MATLAB provides a convenient tool for numerical studies which may enhance understanding and lead to improvements. For example, the MATLAB EIGS program expedites efficient computation of selected eigenvalues of sparse matrices with an Arnoldi iteration [Lehoucq and Sorensen, 1996]. An attempt is made to describe programs in the same order as methods discussed in the previous chapters. Application to low-rank Lyapunov and Sylvester equations is a significant addition. It is here that model-problem ADI iteration has greatest possibilities as a result of the automatic satisfaction of the commutation property lacking in most boundary-value problems. The fact that mathematics initially developed in one context impacts an entirely different application is fortuitous but not uncommon.

6.2 Real Variable: ADIREAL

The general real variable ADI model problem is

$$(HG + VF)\mathbf{u} = \mathbf{s}, \tag{1}$$

[1]The last chapter in the first edition contained a set of FORTRAN programs with sample problems. Since that time I have switched to MATLAB. More economical programs could probably be written in C, C++, or FORTRAN.

E. Wachspress, *The ADI Model Problem*, DOI 10.1007/978-1-4614-5122-8_6,

© Springer Science+Business Media New York 2013

where H and F commute with V and G. (These are expressed as Eqs. 12 and 18 in Chap. 3.) H, V, G, and F are all real SPD matrices of order n. The real variable ADI iteration equations implemented here in the MATLAB program ADIREAL are

$$(H + p_j F)(G\mathbf{u}_{j-1/2}) = \mathbf{s} + (p_j G - V)(F\mathbf{u}_{j-1}), \tag{2.1}$$

$$(V + q_j G)(F\mathbf{u}_j) = \mathbf{s} + (q_j F - H)(G\mathbf{u}_{j-1/2}),$$
$$\text{for } j = 1, 2, \ldots, J. \tag{2.2}$$

The vectors $v_{j-1/2} = G\mathbf{u}_{j-1/2}$ and $\mathbf{v}_j = F\mathbf{u}_j$ are computed during the iteration and the vector $u_J = F^{-1}\mathbf{v}_J$ is the approximation to the solution $\mathbf{u}$. (See Eqs. 26 and 27 in Chap. 3. For the five-point heat diffusion difference equations H and V are tridiagonal for row- and column-ordered equations while F and G are diagonal. For the nine-point finite element equations F and G are tridiagonal.)

The first task is to compute bounds on the eigenvalues of $F^{-1}H$ and $G^{-1}V$. This is easily done by solving the generalized eigenvalue problems $(H - \lambda F)\mathbf{w} = \mathbf{0}$ and $(V - \gamma G)\mathbf{w} = \mathbf{0}$ for minimum and maximum values. The EIGS program provides options for computing selected eigenvalues. The option used here is computation of lower and upper bounds for symmetric matrices. The next task is to align the eigenvalue intervals with the WBJ transformation (Eq. 2 of Chap. 2). This is done with the program WBJREAL. Iteration parameters are then generated to attain a prescribed error bound ε as given in Eqs. 25.1 and 25.2 of Chap. 1. Then the iterations in Eqs. 2.1 and 2.1 are performed with ADITER. All programs were written for serial implementation on a PC. Parallel options are apparent.

```
%ADIREAL yields iteration parameters for
%the real two-variable ADI iteration
H = input('Symmetric matrix H is:');
    if isempty(H)
        nH = input('order of H is:')
            if isempty(nH)
                nH = 100;
            end
        H = triu(ones(nH,nH),-1);
        H = -tril(H,1)+ 3*diag(diag(H));
%Default H is the tridiagonal matrix [-1,2,-1] of
  order 100.
    end
F = input('Positive diagonal matrix F is:');
    if isempty(F)
        F = diag(diag(ones(nH)));
    else
        if length(F) ~= nH
            error('Inconsistent F input.')
        end
    end
```

```
%The default F is the identity matrix.
F = sparse(F); H = sparse(H);
a = eigs(H,F,1,0);
b = eigs(H,F,1);
        nV = input('order of V is:')
            if isempty(nV)
                nV = 50;
            end
V = input('Symmetric matrix V is:');
    if isempty(V)
        V = triu(ones(nV,nV),-1);
        V = -tril(V,1)+ 3*diag(diag(V));
%Default V is the tridiagonal matrix [-1,2,-1] of
  order 50.
    else
        if length(V) ~= nV
            error('Inconsistent V input.')
        end
    end
G = input('Positive diagonal matrix G is:');
    if isempty(G)
        G = diag(diag(ones(nV)));
    else
        if length(G) ~= nV
            error('Inconsistent G input.')
        end
    end
%The default G is the identity matrix.
V = sparse(V); G = sparse(G);
c = eigs(V,G,1,0);
d = eigs(V,G,1);
    if a + c <= 0
        error('Spectra not in positive real plane')
    end
wbjreal
%wbjreal returns the parameters kp,alp,bet,gam,del
%for spectral alignment
    eps = input('Desired error bound is:')
    if isempty(eps)
        eps = 1e-4
    end
q2= eps^2*(1+eps^2/4)^2/16;
qp= kp^2*(1+kp^2/4)^2/16;
J = ceil(.25*log(q2)*log(qp)/pi^2);
rtkp = sqrt(kp);
```

```
ww = zeros(1,J);
    for j = 1:J
        r= (2*j-1)/(2*J);
        nw = 1 + qp^(1-r) + qp^(1+r);
        dw = 1 + qp^r + qp^(2-r);
        xp = (2*r-1)/4;
        qpr = qp^xp;
        ww(j) = rtkp*qpr*nw/dw;
    end
%ww(1:J) are the ADI parameters for the
  aligned spectra.
pj = (alp*ww - bet)./(del - gam*ww);
qj = (alp*ww + bet)./(del + gam*ww);
aditer
return
*****************************************************
*****************************************************
*****************************************************
%wbjreal aligns the spectra for the two sweeps.
    if a+c <= 0
        disp('spectrum not in positive real plane')
        stop
    end
md = (a+c)*(b+d); mn = 2*(b-a)*(d-c);
m = mn/md;
rtm  = sqrt(m*(2+m));
kp = 1/(1+m+rtm);
sig = 2*(a+d)/(b+d);
alp = b*sig - a*(1 + kp);
bet = a*(1+kp) - b*sig*kp;
gam = sig - 1 - kp;
del = 1 + kp - sig*kp;
return
*********************************************
*********************************************
*********************************************
%aditer performs the ADI iteration
    if isempty(usol)
        vsol = zeros(nV,nH);
    else
        vsol = F*usol;
    end
%usol is the initial estimate.
    if isempty(RS)
        RS = ones(nV,nH);
```

```
    end
%RS is the given right hand side.
    for j = 1:J
        rhsj = ((pj(j)*G - V)*vsol+ RS)';
        M = pj(j)*F + H;
        vsol = M\rhsj;
        rhsj = ((qj(j)*F - H)*vsol)' + RS;
        M = qj(j)*G + V;
        vsol = M\rhsj;
    end
usol = F\vsol';
usol = usol';
return
```

6.3 Heat Diffusion Equation: CHEBOUT

Discretization of the heat diffusion equation $-\nabla \cdot D(x,y)\nabla u(x,y) = s(x,y)$ over a rectangular grid may yield either five-point difference equations or nine-point finite element equations as discussed in Chap. 3. This problem with five-point equations is solved with the MATLAB program CHEBOUT. A separable ADI preconditioning inner iteration is accompanied by a Chebyshev outer iteration. ADI iteration parameters are computed with the algorithm described in Eq. 4 of Chap. 1. Theory yields precise iteration parameters to attain a prescribed accuracy.

The program DSAM called by CHEBOUT generates a 16×31 grid with diffusion coefficients on page 55 in 10 × 10 blocks. (Alternatively, one may replace DSAM with RAN5PT to generate a grid of 30 × 30 random diffusion coefficients. Here, a separable preconditioner offers little improvement.) SEPD generates the separable approximation $F * E'$ to D. The program PRECADI calls WBJREAL to align the spectra and determines ADI parameters to attain a prescribed error reduction, eps, with default $eps = 0.01$. CHEBOUT then calls ADITER to perform the iterations. A solution usamp (which may be called with load) accurate to eight significant digits was computed with stringent error bounds. Errors in approximations usol determined with less stringent error bounds may be computed by comparison with usamp.

The Chebyshev outer iteration is standard. Let a and b be the eigenvalue bounds for matrix $B^{-1}A$ where $0 < a < b$ and let $z \equiv \frac{b+a}{b-a}$. The Chebyshev polynomial of the first kind of degree k in z is $T_k(z)$ where $T_0(z) = 1, T_1(z) = z$, and for $k > 1, T_k(z) = 2zT_{k-1}(z) - T_{k-2}(z)$. The k-th approximation to the solution $\mathbf{u}$ of the equation $A\mathbf{u} = \mathbf{s}$ is found with ADI preconditioning $B\mathbf{u}*_k = \mathbf{s} - A\mathbf{u}_{k-1}$ followed by Chebyshev outer iteration with $\mathbf{u}_k = \mathbf{u}_{k-1} + \alpha_k \mathbf{u}*_k + \beta_k[\mathbf{u}_{k-1} - u_{k-2}]$. Here, $\alpha_1 = \frac{2}{a+b}$ and $\beta_1 = 0$. Let $r_k \equiv T_k(z)/T_{k-1}(z)$. We note that $r_1 = z$. The Chebyshev polynomial recursion equation yields $r_k = 2z - \frac{1}{r_{k-1}}$. The extrapolation parameters for $k > 1$ are determined recursively as $\alpha_k = \frac{4}{r_k(a+b)}$ and $\beta_k = \frac{1}{r_k r_{k-1}}$.

```
%chebout is the Chebyshev outer iteration
%for ADI preconditioned heat diffusion
dsam
%The sample problem is the heat diffusion
%equation over
%a 16x31 grid with  x and y increments of unity snd
%diffusion coefficient as on P. 55 in 5x10 blocks.
%The boundary condition is zero value at increments
%of unity from the boundary and symmetric diffusion
%coefficients
%along the grid boundary.
%dsam calls sepd which computes the iteration
%matrices
%Li= H,Gi=F,Lj=V,Gj=G for LiGj + LjGi = RS.
%The separable preconditioner yields an outer
%iteration condition number of ~30 for exact
%ADI solution
bout = max(max(D5./(F5'*E5)));
aout = min(min(D5./(F5'*E5)));
%sam5pt generates the five-point equations for the
 dsam problem.
sam5pt
%ran5pt would generate random diffusion coefficients
 as an alternative precadi
%precadi computes the eigenvalue bounds a,b for
%(H - pF) and c,d, for (V - qG).
%Iteration parameters p(s) and q(s) for
 ADI iterations
%are computed with use of wbjreal for spectral
 alignment.
%The error is reduced by a factor of eps
%so the Chebyshev outer iteration is with:
bout = (1+eps)*bout;
aout = (1-eps)*aout;
z = (bout+aout)/(bout-aout);
 %The spectral radius for iteration without
 %extrapolation is 1/z.
 %The initial estimate is zero,
 %The initial right-hand side is set to unity at all
  grid points
RS = ones(nV,nH);
 %The ADI iterations are performed for the first
  outer iteration.
```

```
            vsol = zeros(nV,nH);
        for j = 1:J
            M = pj(j)*F + H;
            rhsj = (pj(j)*G - V)*vsol + RS;
            vsol = M\rhsj';
            rhsj = ((qj(j)*F - H)*vsol)' + RS;
            M = qj(j)*G + V;
            vsol = M\rhsj;
        end
    usol = F\vsol';
    usol = usol';
    alph = 2/(aout+bout);
    usol = alph*usol;
    delold = usol;
    uold = usol;
    errsol = input('The solution accuracy boundis:')
        if isempty(errsol)
            errsol = 1e-4
        end
    Jout = ceil(acosh(1/errsol)/acosh(z));
    %The error reduction estimate is precise so Jout
    %outer iterations suffice.
    rout = z;
            for jout = 2:Jout
    %The right hand side must be generated for each
     outer iteration
    Auold = zeros(nV,nH);
    for m = 1:nV %row m
        for n = 1:nH %column n
            Auold(m,n) = cP(m,n)*uold(m,n);
            if n < nH
                Auold(m,n) = Auold(m,n)
                                - cE(m,n)*uold(m,n+1);
            end
            if n > 1
                Auold(m,n) = Auold(m,n)
                                - cW(m,n)*uold(m,n-1);
            end
            if m < nV
                Auold(m,n) = Auold(m,n)
                                - cN(m,n)*uold(m+1,n);
            end
```

```
        if m > 1
            Auold(m,n) = Auold(m,n)
                             - cS(m,n)*uold(m-1,n);
        end
    end
end
%The inner iterations are performed:
        vsol = zeros(nV,nH);
        RHS = RS - Auold;
    for j = 1:J
        M = pj(j)*F + H;
        rhsj = (pj(j)*G - V)*vsol+ RHS;
        vsol = M\rhsj';
        rhsj = ((qj(j)*F - H)*vsol)' + RHS;
        M = qj(j)*G + V;
        vsol = M\rhsj;
    end
usol = F\vsol';
delu = usol';
%The Chebyshev outer iteration is performed
rout1 = 2*z - 1/rout;  %rout1 = T_{j+1}/T_j}.
alphout = 4/(rout1*(bout-aout));
betout = 1/(rout1*rout);
rout = rout1;
delu = alphout*delu + betout*delold;
uold = uold + delu;
delold = delu;
        end
disp('The solution is uold:')
uold
disp('The Value of norm(RS-Auold,2)/norm(RS,2) is:')
norm(RS -  Auold,2)/norm(RS,2)
disp('[J = ADI inners per outer  Jout = outers]')
[J Jout]
return
*****************************************************
*****************************************************
*****************************************************
%sam5pt generates 5-point difference equations
%for the sample problem dsam.
%D5(n,m)is ONE HALF the diffusion coefficient in
 quadrant 1 at point (n,m).


D5 = D5/2;
```

```
cN = zeros(nV,nH);cE = zeros(nV,nH);cW = zeros(nV,nH);
 cS = zeros(nV,nH);cP = zeros(nV,nH);
%The diffusion equation coefficients are
%cN, cE, cW,cS, cP where (N,E,W,S = north,east,west
 south and p = center.
%For the interior points:
    for n = 2:nV-1 % row index
        for m = 2:nH-1 % column index
            cN(n,m)  = D5(n,m-1) + D5(n,m);
            cE(n,m)  = D5(n,m) + D5(n-1,m);
            cW(n,m)  = D5(n,m-1) + D5(n-1,m-1);
            cS(n,m)  = D5(n-1,m-1) + D5(n-1,m);
        end
    end
%For the left and right edges:
        for n = 2:nV
            cN(n,1)  = 2*D5(n,1);
            cE(n,1)  = D5(n,1) + D5(n-1,1);
            cW(n,1)  = cE(n,1);
            cS(n,1)  = 2*D5(n-1,1);
            cN(n,nH)  = 2*D5(n,nH-1);
            cE(n,nH)  = D5(n,nH-1) + D5(n-1,nH-1);
            cW(n,nH)= cE(n,nH);
            cS(n,nH)  = 2*D5(n-1,nH);
        end
%For the top and bottom edges:
        for m = 2:nH %col m
            cE(1,m)  = 2*D5(1,m);
            cN(1,m)  = D5(1,m)+ D5(1,m-1);
            cW(1,m)  = 2*D5(1,m-1);
            cS(1,m)  = cN(1,m);
            cE(nV,m)  = 2*D5(nV-1,m);
            cW(nV,m)  = cE(nV,m-1);
            cS(nV,m)  = D5(nV-1,m-1) + D5(nV-1,m);
            cN(nV,m)= cS(nV,m);
        end
%For the corner points (1,1), and (nV,1):
cN(1,1) = cS(2,1); cS(1,1) = cN(1,1);
cS(nV,1) = cN(nV-1,1); cN(nV,1) = cS(nV,1);
cE(1,1) = cW(1,2); cW(1,1) = cE(1,1);
cE(nV,1) = cW(nV,2); cW(nV,1) = cE(nV,1);
%The diagonal coefficient is:
cP = cE + cN + cS + cW;

%All five coefficients have been set for the nV*nH
```

```
 grid points.
D6 = 2*D5; %Half the D5 was used in the coefficient
  calculation.
%Now D6 is the fiffusion coefficient in quadrant 1
  at column m and row n.
D5 = .5*F5'*E5;
%Coefficients are now computed for the separable
 problem.
    for n = 2:nV-1 % row index
        for m = 2:nH-1 % ccolumn index
            ccN(n,m) = D5(n,m-1) + D5(n,m);
            ccE(n,m) = D5(n,m) + D5(n-1,m);
            ccW(n,m) = D5(n,m-1) + D5(n-1,m-1);
            ccS(n,m) = D5(n-1,m-1) + D5(n-1,m);
        end
    end
%for the left and right edges:
        for n = 2:nV
            ccN(n,1) = 2*D5(n,1);
            ccE(n,1) = D5(n,1) + D5(n-1,1);
            ccW(n,1) = ccE(n,1);
            ccS(n,1) = 2*D5(n-1,1);
            ccN(n,nH) = 2*D5(n,nH-1);
            ccE(n,nH) = D5(n,nH-1) + D5(n-1,nH-1);
            ccW(n,nH) = ccE(n,nH);
            ccS(n,nH) = 2*D5(n-1,nH);
        end
%for the top and bottom edges:
        for m = 2:nH %ccol m
            ccE(1,m) = 2*D5(1,m);
            ccN(1,m) = D5(1,m)+ D5(1,m-1);
            ccW(1,m) = 2*D5(1,m-1);
            ccS(1,m) = ccN(1,m);
            ccE(nV,m) = 2*D5(nV-1,m);
            ccW(nV,m) = ccE(nV,m-1);
            ccS(nV,m) = D5(nV-1,m-1) + D5(nV-1,m);
            ccN(nV,m) = ccS(nV,m);
        end
%for the corner points (1,1) and (nV,1):
ccN(1,1) = ccS(2,1); ccS(1,1) = ccN(1,1);
ccS(nV,1) = ccN(nV-1,1); ccN(nV,1) = ccS(nV,1);
ccE(1,1) = ccW(1,2); ccW(1,1) = ccE(1,1);
ccE(nV,1) = ccW(nV,2); ccW(nV,1) = ccE(nV,1);
%The diagonal ccoefficient is:
ccP = ccE + ccN + ccS + ccW;
```

```
%All five separable coefficients have been set for
 the nV*nH grid points.
D5 = 2*D5;
%D5 are the diffusion coefficients for the
 separable problem.
%D6 are the diffusion coefficients for the
 actual problem.
sepd
return
*****************************************************
*****************************************************
*****************************************************
%dsam generates a 16x31 grid of diffusion
 coefficients on page 55 of my ADI book in blocks
 of 5x10.
D5 = ones(16,31);
    for ri = 1:5 %row ri
        for cj = 11:20 %col cj
            D5(ri,cj) = 4;
        end
        for cj = 21:31
            D5(ri,cj) = 36;
%            D5(ri,cj) = 1;
        end
    end
    for ri = 6:10 %row ri
        for cj = 1:10
            D5(ri,cj) = 16;
%            D5(ri,cj) = 4;
        end
        for cj = 11:20 %col cj
            D5(ri,cj) = 100;
%             D5(ri,cj)=8;
        end
        for cj = 21:31
            D5(ri,cj) = 1600;
%            D5(ri,cj) = 4;
        end
    end
    for ri = 11:16 %row ri
        for cj = 1:10
            D5(ri,cj) = 9;
%            D5(ri,cj) = 1;
        end
        for cj = 11:20 %col cj
```

```
            D5(ri,cj) = 25;
%             D5(ri,cj) = 4;
        end
    end
nV = 16; nH = 31;
%D5 is the sample problem diffusion coefficient.
%The boundary condition is zero on columns = 0,nH+1
  and rows = 0, nV+1.
sepd
return
****************************************************
****************************************************
****************************************************
%ran5pt generates coefficients for 5-point
%difference equations with Diff coef=2*(minel+rand).
nH = input('number of columns nH =');
if isempty(nH)
    nH = 31;
end
nV = input('number of rows nV=');
if isempty(nV)
    nV = 16;
end
minel = input('minel is lower bound on elements');
if isempty(minel)
    minel = .001;
end
D5 = minel + rand(nV,nH);
%D5(n,m)is ONE HALF the diffusion coefficient
 in quadrant 1 at point (n,m).
cN = zeros(nV,nH);cE = zeros(nV,nH);cW = zeros(nV,nH);
 cS = zeros(nV,nH);cP = zeros(nV,nH);
%The diffusion equation coefficients are
%cN, cE, cW,cS, cP where
%for the interioe points:
    for n = 2:nV-1 %row n
        for m = 2:nH-1 %column m
            cN(n,m) = D5(n,m-1) + D5(n,m);
            cE(n,m) = D5(n,m) + D5(n-1,m);
            cW(n,m) = D5(n,m-1) + D5(n-1,m-1);
            cS(n,m) = D5(n-1,m-1) + D5(n-1,m);
        end
    end
%for the left and right edges:
        for n = 2:nV
```

```
                cN(n,1)  = 2*D5(n,1);
                cE(n,1)  = D5(n,1) + D5(n-1,1);
                cW(n,1)  = cE(n,1);
                cS(n,1)  = 2*D5(n-1,1);
                cN(n,nH)  = 2*D5(n,nH-1);
                cE(n,nH)  = D5(n,nH-1) + D5(n-1,nH-1);
                cW(n,nH)  = cE(n,nH);
                cS(n,nH)  = 2*D5(n-1,nH);
        end
%for the top and bottome edges:
        for m = 2:nH
                cE(1,m)  = 2*D5(1,m);
                cN(1,m)  = D5(1,m)+ D5(1,m-1);
                cW(1,m)  = 2*D5(1,m-1);
                cS(1,m)  = cN(1,m);
                cE(nV,m)  = 2*D5(nV-1,m);
                cW(nV,m)  = cE(nV,m-1);
                cS(nV,m)  = D5(nV-1,m-1) + D5(nV-1,m);
                cN(nV,m)  = cS(nV,m);
        end
%for the four corner points:
cN(1,1)  = cS(2,1); cS(1,1)  = cN(1,1);
 cN(1,nH)  = cS(2,nH);
cS(1,nH)  = cN(1,nH); cS(nV,1)  = cN(nV-1,1);
 cN(nV,1)  = cS(nV,1);
cE(1,1)  = cW(1,2); cW(1,1)  = cE(1,1);
cE(nV,1)  = cW(nV,2);
cW(nV,1)  = cE(nV,1);
%The diagonal coefficient is:
cP = cE + cN + cS + cW;
%All five coefficients have been set for the nV*nH
 grid points.
D5 = 2*D5; %Half the D5 was used in the coefficient
 calculation.
%Now D5 is the fiffusion coefficient in quadrant 1
 at column n and row m.
D6 = D5;
sepd
return
********************************************
********************************************
********************************************

%sepd computes E5 and F5 to approximate D5
%with F5'*E5.
```

```
%The number of columns is nH and rows is nV.
%The tridiagonal Li and Lj matrices are generated.
%Program generated on 1/2/2012.
E5 = ones(1,nH);
Li = zeros(nH);
Lj = zeros(nV);
Flast = ones(1,nV);
r1 = ones(1,nV);
c1 = ones(1,nH);
Drat = ones(nV,nH);
F5 = Flast;
k = 1;
    for j = 1:nV %row j
        cj1 = max(D5(j,:)); cj2 = min(D5(j,:));
        F5(j) = sqrt(cj1*cj2);
    end
frat = 1;
            while frat > 0.001
    for i = 1:nH %col i
        for j = 1:nV %row j
            r1(j) = D5(j,i)/F5(j);
        end
        ri1 = max(r1); ri2 = min(r1);
        E5(i) = sqrt(ri1*ri2);
    end
    for j = 1:nV %row j
        for i = 1:nH %col i
            c1(i) = D5(j,i)/E5(i);
        end
        cj1 = max(c1); cj2 = min(c1);
        F5(j) = sqrt(cj1*cj2);
    end
    frat = norm(Flast - F5)/norm(F5);
    Flast = F5;
            end
              for i = 1:nH %col i
                for j = 1:nV %row j
                  Drat(j,i) = D5(j,i)/(E5(i)*F5(j));
                    end
                end
maxrat = max(max(Drat));
minrat = min(min(Drat));

pprec = maxrat/minrat;
mamD5 = max(max(D5)); minD5 = min(min(D5));
```

```
pD5 = mamD5/minD5;
benef = pD5/pprec;
    for ic = 2:nH-1
        Li(ic,ic+1) = -E5(ic);
        Li(ic,ic-1) = -E5(ic-1);
        Li(ic,ic) = E5(ic) + E5(ic-1);
    end
    Li(1,2) = -E5(1); Li(1,1) = 2*E5(1);
    Li(nH,nH) = E5(nH) + E5(nH-1);
    Li(nH,nH-1) = -E5(nH-1);
H = sparse(Li);
F = sparse(diag(0.5*diag(H)));
    for jc = 2:nV-1
        Lj(jc,jc+1) = -F5(jc);
        Lj(jc,jc-1) = -F5(jc-1);
        Lj(jc,jc) = F5(jc) + F5(jc-1);
    end
    Lj(1,2) = -F5(1); Lj(1,1) = 2*F5(1);
    Lj(nV,nV) = F5(nV) + F5(nV-1);
    Lj(nV,nV-1) = -F5(nV-1);
V = sparse(Lj);
G = sparse(diag(0.5*diag(V)));
return
**************************************
**************************************
**************************************
%precadi generates the precondition ADI pj and qj.
%H,F,V, and G are given as sparse matrices.
nH = length(H); nV = length(V);
opts.tol = 1e-6;
opts.disp = 0;
a = eigs(H,F,1,'sa',opts);
b = eigs(H,F,1,'la',opts);
c = eigs(V,G,1,'sa',opts);
d = eigs(V,G,1,'la',opts);
    if a + c <= 0
        error('Spectra not in positive real plane')
    end
wbjreal
%wbjreal returns the parameters kp,alp,bet,gam,del
%for spectral alignment


    eps = input('Desired error bound is:')
    if isempty(eps)
```

```
            eps = 0.01
        end
    q2= eps^2*(1+eps^2/4)^2/16;
    qp= kp^2*(1+kp^2/4)^2/16;
    J = ceil(0.25*log(q2)*log(qp)/pi^2);
    rtkp = sqrt(kp);
    ww = zeros(1,J);
        for j = 1:J
            r= (2*j-1)/(2*J);
            nw = 1 + qp^(1-r) + qp^(1+r);
            dw = 1 + qp^r + qp^(2-r);
            xp = (2*r-1)/4;
            qpr = qp^xp;
            ww(j) = rtkp*qpr*nw/dw;
        end
    %ww(1:J)are theADI parametersforthealignedspectra.
    pj = (alp*ww - bet)./(del - gam*ww);
    qj = (alp*ww + bet)./(del + gam*ww);
    return
```

6.4 Similarity Reduction to Banded Form: QGBAND

A nonsymmetric matrix A is reduced with QGBAND by the method described in Chap. 5 to the banded upper Hessenberg matrix $S = \text{snorm}\, GAG^{-1}$. The Lyapunov matrix equation $AX + XA^\top = C$ is reduced with QGLAP to $SX + XS^\top = CS$. Matrices S, G^{-1}, and $CS = GCG^\top$ are computed. The Sylvester equation, $AX + XB = C$, is reduced with QGSYL to $SX + XB = CS$. Matrices $S = \text{snorm}\, GAG^{-1}$, $SB = \text{snorm}\, HBH^{-1}$, G^{-1}, and H are saved. The low-rank Lyapunov equation with $C = cl * cl'$ is solved with QGLOW. Now matrices $S, G * cl$ and G^{-1} are computed.

```
%QGBAND computes a banded upper Hessenberg matrix S
 similar to %snorm*A.
%The last nonzero element in row k is element k
 of sband.
%Columns are reduced successively with Householder
 transformations.
%After column k is reduced, row k1 (the last
 unreduced row beyond column
%k+1)is reduced with a householder transformation
 beyond column k+2.
%Element S(k1,k+2) then has magnitude
%equal to the 2-norm of S(k1,k+2:n) prior to
```

```
 reduction. If
%|S(k1,k+2)/S(k1,k+1)| < tol, S(k1,k+2) is reduced to
 zero with a gaussian
%transformation and k1 is increased to k1+1. 10/31/11
A = input('Matrix A is:');
if isempty(A)
    A = randn(30);
end
myzero = 1e-10;
n=length(A); nA = n;
snorm = 1/sqrt(norm(A,1)*norm(A,inf));
k1 = 1; S = snorm*A;
%This normalizes S so its eigenvalue magnitudes are
 bounded by unity.
band = linspace(n,n,n);
tol = input('Bound on gaussian multipliers is:');
if isempty(tol)
    tol = 1e3
end
%The 1-norm of S is initially around 1.  It has been
 observed that
%The 1-norm of the banded upper Hessenberg matrix
 similar to the initial S
%is of order magnitude less than tol.  Eigenvalues of
the reduced matrix are
%reasonably close to those of S even when tol
 is large.
                    for k = 1:n-2
kp1 = k+1; kp2 = k+2;kp3 = min(k+3,n);
colvec = S(kp1:n,k); Y = max(abs(colvec));
%The column is reduced with a HH transformation:
vk = S(kp1:n,k);
%MATLAB sgn(0) = 0 and we need sgn(0)=1 here so:
    if vk(1) == 0
        sgn = 1;
    else
        sgn = sign(vk(1));
    end
u1 = sgn*norm(vk,2);
uk = [u1;zeros(n-kp1,1)]; wk = uk + vk; nmk = wk'*wk;
HHk = eye(n-k) - 2*(wk*wk')/nmk;
S(kp1:n,k:n) = HHk*S(kp1:n,k:n);
S(k1:n,kp1:n) = S(k1:n,kp1:n)*HHk;
S(kp2:n,k) = zeros(n-kp1,1); %To eliminate roundoff
 error.
```

```
%The column has been reduced.
%The matrix is banded above k.
rowvec = S(k1,kp2:n); Z = max(abs(rowvec));
                                    if k < n-2
    if Z < myzero
%The row does not have to be reduced.
        S(k1,kp2:n) = zeros(1,n-kp1);
        band(k1) = kp1;
        k1 = k1+1;
    else
%Row k1 is reduced beyond kp2 with a HH
 transformation.
vk = S(k1,kp2:n);
    if vk(1) == 0
        sgn = 1
    else
        sgn = sign(vk(1));
    end
u1 = sgn*norm(vk,2);
uk = [u1 zeros(1,n-kp2)]; wk = uk + vk; nmk = wk*wk';
HHk = eye(n-kp1) - 2*(wk'*wk)/nmk;
S(k1:n,kp2:n) = S(k1:n,kp2:n)*HHk;
S(kp2:n,kp1:n) = HHk*S(kp2:n,kp1:n);
S(k1,kp3:n) = zeros(1,n-kp3+1); %To eliminate
 roundoff error.
band(k1) = kp2;
    end
                                    end %on k < n-2
      if Z > myzero
         dm = S(k1,kp1);
         if abs(dm) > sqrt(myzero)
%We need a significant dm to attempt row trduction.
tolk = abs(S(k1,kp2)/dm);
          if tolk < tol
%We reduce S(k1,kp2) to zero with a gaussian
 transformation.
%Else element S(k1,kp2) cannot be reduced with
 a gaussian transformation.
w = S(k1,kp2)/dm;
S(k1+1:n,kp2) = S(k1+1:n,kp2) - w*S(k1+1:n,kp1);

S(k1,kp2) = 0;
S(kp1,kp1:n) = S(kp1,kp1:n) + w*S(kp2,kp1:n);
band(k1) = kp1; k1 = k1+1;
          end
```

```
            end
        end %on Z > myzero
          end %on k =1:n-2
k = n-1;
if abs(S(n-2,n))< myzero
    S(n-2,n) = 0;
    band(n-2) = n-1;
end
sband = band;
%*****S is banded upper Hessenberg*****
disp('1-norm of S is')
norm(S,1)
figure
spy(S);
hold on
title('Figure 1: Profile of matrix S')
ylabel('matrix row'); xlabel('nonzero elements')
hold off
disp('number of elements beyond tridiagonal')
xtra = nnz(triu(S,2))
eigon = input('To compute eigenvalues enter 1,
  else enter 0')
    if eigon == 1
%All eigenvalues of S are found with MATLAB sparse
  matrix routine eigs.
%My MATLAB eigs does not permit eigs(T,nA) so:
T = sparse(S);
lam0 = eigs(T,2,0);
lamT = eigs(T,nA-2);
lamS = [lam0;lamT];
lamS = sort(lamS);
%If the second element is complex it may happen that
 the third
%element which should be the conjugate has the same
 imaginary part.
%This coujld be resolved but does not affect
 my application.

eigsofA = lamS/snorm;
disp('eigenvalues of A are in eigsofA')
    end
return
*********************************************
*********************************************
*********************************************
```

```
%QGLAP computes a banded upper Hessenberg matrix S
 similar to %snorm*A.
%The last nonzero element in row k is element k
 of sband.
%Columns are reduced successively with Householder
 transformations.
%After column k is reduced, row k1 (the last
 unreduced row beyond column
%k+1)is reduced with a householder transformation
 beyond column k+2.
%Element S(k1,k+2) then has magnitude
%equal to the 2-norm of S(k1,k+2:n) prior to
 reduction. If
%|S(k1,k+2)/S(k1,k+1)| < tol, S(k1,k+2) is reduced to
 zero with a gaussian
%transformation and k1 is increased to k1+1.
%A file GI is also generated for recovery%
 of X = GI*Y*GI^T.
%AX +XA' = C is solved for X by calling qgpar
 and adilap.
%Eigenvalues of A must lie in the right half plane
 bounded away
%from the imaginary axis and C must be SPD.
A = input('Matrix A is:');
if isempty(A)
    randn('seed',0)
    A = randn(30) + 5.5*eye(30);;
end
nA = length(A);
Xtr = zeros(nA);
C = input('matrix C is:');
if isempty(C)
    C = rand(nA); C = C+C';
    c1 = min(eig(C));
    C = C + (.1+abs(c1))*eye(nA);
%This yields an spd C.
end

myzero = 1e-14;
n = nA; snorm = 1/sqrt(norm(A,1)*norm(A,inf));
k1 = 1;
S = snorm*A;
%This normalizes S so its eigenvalue magnitudes
 are bounded by unity.
CS = snorm*C;
```

```
GI = eye(n);
band = linspace(n,n,n);
tol = input('Bound on gaussian multipliers is:');
if isempty(tol)
    tol = 1e3
end
%The 1-norm of S is initially around 1.
 It has been observed that
%The 1-norm of the banded upper Hessenberg matrix
 similar to the initial S
%is of order magnitude less than tol.
 Eigenvalues of the reduced matrix are
%reasonably close to those of S even when
 tol is large.
                          for k = 1:n-2
kp1 = k+1; kp2 = k+2;kp3 = min(k+3,n);
colvec = S(kp1:n,k); Y = max(abs(colvec));
%The column is reduced with a HH transformation:
vk = S(kp1:n,k);
    if vk(1) == 0
        sgn = 1
    else
        sgn = sign(vk(1));
    end
u1 = sgn*norm(vk,2);
uk = [u1;zeros(n-kp1,1)]; wk = uk + vk; nmk = wk'*wk;
HHk = eye(n-k) - 2*(wk*wk')/nmk;
S(kp1:n,k:n) = HHk*S(kp1:n,k:n);
CS(kp1:n,:) = HHk*CS(kp1:n,:);
S(k1:n,kp1:n) = S(k1:n,kp1:n)*HHk;
CS(:,kp1:n) = CS(:,kp1:n)*HHk;
S(kp2:n,k) = zeros(n-kp1,1); %To eliminate
  roundoff error.
%The column has been reduced.
%The matrix is banded above k.
GI(:,kp1:n) = GI(:,kp1:n)*HHk;
rowvec = S(k1,kp1:n); Z = max(abs(rowvec));
                                  if k < n-2
nm = norm(S(k1,kp2:n),inf);
    if nm < myzero
%The row does not have to be reduced.
        S(k1,kp2:n) = zeros(1,n-kp1);
        k1 = k1+1; band(k1) = kp1;
    else
%Row k1 is reduced beyond kp2 with a HH
```

```
  %transformation .
vk = S(k1,kp2:n);
    if vk(1) == 0
        sgn = 1
    else
        sgn = sign(vk(1));
    end
u1 = sgn*norm(vk,2);
uk = [u1 zeros(1,n-kp2)]; wk = uk + vk; nmk = wk*wk';
HHk = eye(n-kp1) - 2*(wk'*wk)/nmk;
S(k1:n,kp2:n) = S(k1:n,kp2:n)*HHk;
CS(:,kp2:n) = CS(:,kp2:n)*HHk;
GI(:,kp2:n) = GI(:,kp2:n)*HHk;
S(kp2:n,kp1:n) = HHk*S(kp2:n,kp1:n);
CS(kp2:n,:) = HHk*CS(kp2:n,:);
S(k1,kp3:n) = zeros(1,n-kp3+1); %To eliminate
 roundoff error.
band(k1) = kp2;
dm = S(k1,kp1);
            if abs(dm) > sqrt(myzero)
tolk = abs(S(k1,kp2)/dm);
             if tolk > tol
%Element S(k1,kp2) cannot be reduced with a
 gaussian transformation.
             else
%We reduce S(k1,kp2) to zero with a
 gaussian transformation.
w = S(k1,kp2)/dm;
S(k1+1:n,kp2) = S(k1+1:n,kp2) - w*S(k1+1:n,kp1);
CS(:,kp1) = CS(:,kp1) + w*CS(:,kp2);
S(k1,kp2) = 0;
S(kp1,kp1:n) = S(kp1,kp1:n) + w*S(kp2,kp1:n);
CS(kp1,:) = CS(kp1,:) + w*CS(kp2,:);
GI(:,kp2) = GI(:,kp2) - w*GI(:,kp1);
band(k1) = kp1; k1 = k1+1;
             end
            end
    end
                           end %on k < n-2
                   end %on k =1:n-2
        if k1 == n-2
dm = S(k1,kp1);
            if abs(dm) > sqrt(myzero)
tolk = abs(S(k1,kp2)/dm);
             if tolk > tol
```

```
%Element S(k1,kp2) cannot be reduced with a
 gaussian transformation.
          else
%We reduce S(k1,kp2) to zero with a gaussian
 transformation.
w = S(k1,kp2)/dm;
S(k1+1:n,kp2) = S(k1+1:n,kp2) - w*S(k1+1:n,kp1);
CS(:,kp1) = CS(:,kp1) + w*CS(:,kp2);
S(k1,kp2) = 0;
S(kp1,kp1:n) = S(kp1,kp1:n) + w*S(kp2,kp1:n);
CS(kp1,:) = CS(kp1,:) + w*CS(kp2,:);
GI(:,kp2) = GI(:,kp2) - w*GI(:,kp1);
band(k1) = kp1; k1 = k1+1;
          end
         end
      end
k = n-1;
sband = band;
%*****S is banded upper Hessenberg*****
disp('1-norm of S is')
norm(S,1)
figure
spy(S);
hold on
title('Figure 1:  Profile of matrix S')
ylabel('matrix row'); xlabel('nonzero elements')
hold off
disp('number of elements beyond tridiagonal')
xtra = nnz(triu(S,2))
%ADI iteration parameters are determined:
errY = input('allowed error in Y')
    if isempty(errY)
        errY = 1e-6
    end

qgpar
usol = zeros(n);
adilap
    if Yerr > 10*errY
%Another set of ADI iterations is performed.
        adilap
    end
X = GI*usol*GI';
disp('The estimated value for ||error in X||/||X||
 is')
```

```
norm(C-A*X-X*A',1)/norm(C,1)
    if Xtr ~= zeros(nA)
disp('The true value for ||error in X||/||X|| is')
norm(X - Xtr,1)/norm(Xtr,1)
    end
return
***********************************************
***********************************************
***********************************************
```

6.5 Solution of the Lyapunov Equation with QGLAP

S is stored as a sparse matrix of order n. For ADI iterative solution of Lyapunov equations crucial values are either near the imaginary axis or of large magnitude. Let $nlamin = \text{ceil}(2\sqrt{n})$ and $nlamax = \text{ceil}(\sqrt{nlamin})$. The crucial eigenvalues are computed in QGPAR with the commands $lamin = eigs(S, nlamin, sr)$ and $lamax = eigs(S, nlamax)$. (Initially $nlamin$ was chosen as ceil$\sqrt{n}$. However, in one trial case, a crucial value was missed and convergence was poor. One could modify the program to recognize this and iterate with new parameters. It is hoped that the default value will suffice in the vast majority of cases.)

Eigenvalues subtending angles greater than 1 rad at the origin are treated distinctly. The remaining eigenvalues are embedded in an elliptic function region and iteration parameters to attain a prescribed error reduction of ε are computed as described in Chap. 4. ADILAP iterates with parameters determined in QGPAR The default choice of $C = A + A^{\mathsf{T}}$ yields the identity matrix as a solution so that a norm of the approximation error may be computed. In the absence of truth, the error in the computed X is estimated reasonably well by $||C - AX - XA^{\mathsf{T}}||/||C||$.

```
%qgpar computes the selected ceil(sqrt(n)) small
 eigenvalues
%and ceil(n^1/4)) large eigenvalues of S.
%ADI iteration parameters to attain a desired error
 reduction
%are then generated.
T = sparse(S);
nlamin = ceil(2*sqrt(n));
nlamax = ceil(sqrt(nlamin));
opts.tol = 1e-6;
opts.disp = 0;
lamin = eigs(T,nlamin,'sr',opts);
lamax = eigs(T,nlamax,'lm',opts);
j1 = 0;
wmag = 0;
if min(real(lamin)) <= 0
```

```
disp('Spectrum not in positive real half plane.');
disp('Hit [enter] to continue anyway
 and [ctrl-c] to end.')
pause
end
%We choose a conservative bound on the
 spectral radius:
b = 3*abs(lamax(1));
ang = angle(lamin);
angmax = max(ang);
ka = 1;
            if angmax >= pi/3
%Eigenvalues at angles greater than 60 deg are chosen
%as iteration parameters and excluded from
%eigenvalues enclosed by the elliptic function region.
%This assures only real parameters for the
%residual regions.
    for j = 1:nlamin
        if ang(j) > pi/3
            ww(ka) = lamin(j);
            ww(ka+1) = conj(lamin(j));
            ka = ka+2;
            lamin(j) = b;
        end
    end
angmax = max(angle(lamin));
            end
rat = real(lamin(1))/b;
if rat < 1e-6
  disp('Poorly conditioned problem with eig ratio=')
rat
  disp('Hit [enter] to continue and [ctrl-c]
    to stop.')
pause
end
j = 1;
    while rat < 1e-6
        ww(ka) = lamin(j);
        lamin(j) = b;
        ka = ka + 1;
        j = j + 1;
        rat = abs(lamin(j))/b;
    end
%We choose a conservative lower bound on the
 real components:
```

```
a = 0.8*min(real(lamin));
ap = a;
bp = b;
rtab = sqrt(a*b);
    j1 = ka-1;
%We normalize ww:
    for j = 1:j1
       ww(j) = ww(j)/rtab;
    end
csqa = cos(angmax)^2;
csqb = 4/(2+a/b+b/a);
mm = 2*csqa/csqb - 1;
%By choosing angmax < pi/3, we ensure mm > 1.
%For larger angles admitted.*************************
    if mm < 1
%Complex parameters must be computed
%from a dual spectrum.
x1 = lamin/rtab;
x2 = rtab./lamin;
x12 = [x1;x2];
x3 = lamax/rtab;
x4 = rtab./lamax;
x34 = [x3;x4];
x13 = [x1;x3];
x24 = [x2;x4];
xx = 2./(x13 + x24);
z1 = xx + sqrt(xx.*xx - 1);
z2 = xx - sqrt(xx.*xx - 1);
z = [z1;z2];
z = real(z) + i*abs(imag(z));
mm = 2*csqb/csqa - 1;
ap = tan(pi/4 - angmax/2);
bp = 1/ap;
wmag = 1
    end
%Resume with mm >= 1********************************
%The error reduction errY now determines J.
kp = mm + sqrt(mm^2-1);
kp = 1/kp;
kp2 = kp^2;
k = sqrt(1-kp2);
zz = (1 - sqrt(k))/(2*(1+sqrt(k)));
    if zz^2 <0 .5
       qp = zz*(1+zz^4);
       q = exp(pi^2/log(qp));
```

```
     else
          zp = sqrt(1-zz^2);
          q = zp*(1+zp^4);
          qp = exp(pi^2/log(q));
     end
          if angmax <0 .01
%When the angle is < 0.01 we compute the
%ADI error reduction for a real spectrum.
     Jl = ceil(0.5*log(errY/40)/log(q));
          else
%Since MATLAB ellipj computes elliptic integrals
%we use sn(z) = sin(phi) and F(phi) to compute vv.
rtkp = sqrt(kp);
sig = ap/(bp*kp);
fl = 1 - ap*kp/b;
el = 1 - ap/(bp*kp);
snrK = sqrt(el/fl);
ph0 = asin(snrK);
Kp = ellipke(kp2);
cagm
y = 0.5*vv/Kp;
vJ = 1 - 2*y;
Jl = ceil(log(errY/40)/(2*vJ*log(q)));
          end
     if j1 ~= 0
          ww = [zeros(1,Jl) ww(1:j1)];
     else
          ww = zeros(1,Jl);
     end
for j = 1:Jl
     rr = (2*j-1)/(2*Jl);
     qexp = (2*rr-1)/4;

     numer = 1 + qp^(1-rr) + qp^(1+rr);
     denom = 1 + qp^rr + qp^(2-rr);
     ww(j) = qp^qexp*numer/denom;
end
ww(1:Jl) = ww(1:Jl)/sqrt(ww(1)*ww(Jl));
          if wmag == 0 %Always for angmax < pi/3.
%We now denormalize ww:
               ww = rtab*ww;
          else
     JJ = floor(Jl/2);
     for j = 1:JJ
      par = 2/(ww(j+j1) + 1/ww(j+j1));
```

```
         angj = acos(par);
         wwp(j) = rtab*exp(i*angj);
        end
        if j1 ~= 0
            ww(1:j1) = rtab*ww(1:j1);
        end
                for j = 1:JJ
                    ww(2*j-1) = wwp(j);
                    ww(2*j) = wwp(j)';
                end
              if Jl/2 ~= JJ
                ww(Jl) = rtab;
              end
            end
    J = Jl + j1;
    %The iteration parameters for ADI error
    %reduction [terror] are the J values in ww(1:J)
    %with the j1 discrete complex values at the end.
    disp('qgpar ended with j1 and J = ')
    j1
    J
    return
    *******************************************
    *******************************************
    *******************************************
    %adilap performs the ADI iteration to solve
    %the reduced Lyapunov equation.
        if isempty(usol)
            usol = zeros(n);
        end

    %usol is the initial estimate.
        if isempty(CS)
            error('no source term')
        end
            Tp = T';
        for j = 1:J
            rhsj = ww(j)*usol - usol*Tp+ CS;
            M = ww(j)*speye(n) + T;
            usol = M\rhsj;
            usp = usol';
            rhsj = ww(j)*usp - usp*Tp + CS;
            usol = (M\rhsj)';
        end
    %usol is now Y.
```

```
Cest = T*usol + usol*T';
disp('Estimated ||error in Y||/||Y||')
Yerr = norm(Cest - CS,1)/norm(CS,1)
return
****************************************
****************************************
****************************************
%cagm is the AGM algoirthm for complex
elliptic regions
%Given ph0 and kp.
k = sqrt(1 - kp^2);
Kp = ellipke(kp^2);
aa = 1; bb = k;cc = 1;
ncag = 1;
phi = ph0;
            while cc > 0.0001
adp = atan(bb*tan(phi)/aa);
phig0 = phi + adp;
        if phig0 < 2*phi
    while abs(phig0 - 2*phi) > abs(phig0 + pi-2*phi)
        phig0 = phig0 + pi;
    end
        else
    while abs(phig0 - 2*phi) > abs(phig0 - pi-2*phi)
        phig0 = phig0 - pi;
    end
        end

phi = phig0;
        aa = (aa+bb)/2; bb = sqrt(aa*bb);
        cc = 0.5*(aa - bb);
        ncag = ncag + 1;
            end
dd = 2^(ncag-2)*pi;
vv = phi/(aa*dd);
return
```

6.6 Solution of the Sylvester Equation with QGSYL

QGSYL converts S and SB to sparse matrices T and TB and eigenvalue bounds for each matrix are computed with $eigs$ as in QGPAR. WBJREAL is then called to align the real intercepts to $[k', 1]$. The spectra are then normalized to $(\sqrt{k'}, 1/\sqrt{k'})$. Eigenvalues which subtend angles greater than 1 rad at the origin are handled distinctly. Complex residual spectra are aligned as described in Sect. 6.4 in Chap. 4. Transformed parameters are computed for the aligned spectra and actual parameters by back transformation from both the complex alignment and the WBJREAL alignment. The iteration is performed with ADISYL.

The back transformation of aligned elliptic function regions leads to iteration parameters which do not necessarily fall within the actual elliptic function regions. Parameters with large negative real components were sometimes generated. The iteration with these parameters converged as anticipated.

```
%qgsyl solves the Sylvester matrix equation with
%a full right-hand side.
A = input('matrix A is:');
     if isempty(A)
         randn('seed',0)
         A = randn(40) + 6.5*eye(40);
     end
nA = length(A);
B = input('matrix B = ');
     if isempty(B)
         randn('seed',0)
         B = randn(20) + 4.2*eye(20);
     end
mB = length(B);
Xtr = zeros(nA,mB);
C = input('Matrix C = ');
     if isempty(C)
         Xtr = rand(nA,mB);
         C = A*Xtr + Xtr*B;
%This sets the solution to Xtr.
     end
   if size(C) ~= [nA mB]
         error('C wrong dimension.')
         return
   end
%We now compite the banded upper-Hessenberg matrix
%S similar to A:
myzero = 1e-14;
n = nA;
snorm = 1/sqrt(norm(A,1)*norm(A,inf));
tnorm = 1/sqrt(norm(B,1)*norm(B,inf));
snorm = sqrt(snorm*tnorm);
k1 = 1;
S = snorm*A;
CS = snorm*C;
GI = eye(n);
```

```
tol = input('Bound on gaussian multipliers is:');
if isempty(tol)
    tol = 1e3
end
                                for k = 1:n-2
kp1 = k+1; kp2 = k+2;kp3 = min(k+3,n);
colvec = S(kp1:n,k); Y = max(abs(colvec));
%The column is reduced with a HH transformation:
vk = S(kp1:n,k);
    if vk(1) == 0
        sgn = 1
    else
        sgn = sign(vk(1));
    end
u1 = sgn*norm(vk,2);
uk = [u1;zeros(n-kp1,1)]; wk = uk + vk; nmk = wk'*wk;
HHk = eye(n-k) - 2*(wk*wk')/nmk;
S(kp1:n,k:n) = HHk*S(kp1:n,k:n);
CS(kp1:n,:) = HHk*CS(kp1:n,:);
S(k1:n,kp1:n) = S(k1:n,kp1:n)*HHk;
S(kp2:n,k) = zeros(n-kp1,1); %To eliminate
 %roundoff error.
%The column has been reduced.
%The matrix is banded above k.
GI(:,kp1:n) = GI(:,kp1:n)*HHk;
rowvec = S(k1,kp1:n); Z = max(abs(rowvec));
                                          if k < n-2
nm = norm(S(k1,kp2:n),inf);
    if nm < myzero
%The row does not have to be reduced.
        S(k1,kp2:n) = zeros(1,n-kp1);
        k1 = k1+1;
    else
%Row k1 is reduced beyond kp2 with a HH transformation.
vk = S(k1,kp2:n);
    if vk(1) == 0
        sgn = 1
    else
        sgn = sign(vk(1));
    end
u1 = sgn*norm(vk,2);
uk = [u1 zeros(1,n-kp2)]; wk = uk + vk; nmk = wk*wk';
HHk = eye(n-kp1) - 2*(wk'*wk)/nmk;
S(k1:n,kp2:n) = S(k1:n,kp2:n)*HHk;
GI(:,kp2:n) = GI(:,kp2:n)*HHk;
```

```
S(kp2:n,kp1:n) = HHk*S(kp2:n,kp1:n);
CS(kp2:n,:) = HHk*CS(kp2:n,:);
S(k1,kp3:n) = zeros(1,n-kp3+1); %To eliminate
 roundoff error.
dm = S(k1,kp1);
           if abs(dm) > sqrt(myzero)
tolk = abs(S(k1,kp2)/dm);
            if tolk > tol
%Element S(k1,kp2) cannot be reduced with
 a gaussian transformation.
            else
%We reduce S(k1,kp2) to zero with a gaussian
 transformation.
w = S(k1,kp2)/dm;
S(k1+1:n,kp2) = S(k1+1:n,kp2) - w*S(k1+1:n,kp1);
S(k1,kp2) = 0;
S(kp1,kp1:n) = S(kp1,kp1:n) + w*S(kp2,kp1:n);
CS(kp1,:) = CS(kp1,:) + w*CS(kp2,:);
GI(:,kp2) = GI(:,kp2) - w*GI(:,kp1);
k1 = k1+1;
            end
           end
    end
                        end %on k < n-2
                end %on k =1:n-2
         if k1 == n-2
dm = S(k1,kp1);

           if abs(dm) > sqrt(myzero)
tolk = abs(S(k1,kp2)/dm);
            if tolk > tol
%Element S(k1,kp2) cannot be reduced with%
 a gaussian transformation.
            else
%We reduce S(k1,kp2) to zero with
 a gaussian transformation.
w = S(k1,kp2)/dm;
S(k1+1:n,kp2) = S(k1+1:n,kp2) - w*S(k1+1:n,kp1);
S(k1,kp2) = 0;
S(kp1,kp1:n) = S(kp1,kp1:n) + w*S(kp2,kp1:n);
CS(kp1,:) = CS(kp1,:) + w*CS(kp2,:);
GI(:,kp2)= GI(:,kp2) - w*GI(:,kp1);
k1 = k1+1;
            end
           end
```

```
        end
k = n-1;
%S is snorm*G*A*G^-1, CS = snorm*G*C, GI is G^-1.
disp('1-norm of S is')
norm(S,1)
disp('number of elements beyond tridiagonal')
xtra = nnz(triu(S,2))
%We now reduce B to banded upper Hessenberg form SB.
m = mB;
k1 = 1;
H = eye(m);
SB = snorm*B;
                    for k = 1:m-2
kp1 = k+1; kp2 = k+2;kp3 = min(k+3,m);
colvec = SB(kp1:m,k); Y = max(abs(colvec));
%The column is reduced with a HH transformation:
vk = SB(kp1:m,k);
    if vk(1) == 0
        sgn = 1
    else
        sgn = sign(vk(1));
    end
u1 = sgn*norm(vk,2);
uk = [u1;zeros(m-kp1,1)]; wk = uk + vk; nmk = wk'*wk;
HHk = eye(m-k) - 2*(wk*wk')/nmk;

SB(kp1:m,k:m) = HHk*SB(kp1:m,k:m);
SB(k1:m,kp1:m) = SB(k1:m,kp1:m)*HHk;
SB(kp2:m,k) = zeros(m-kp1,1); %To eliminate
%roundoff error.
CS(:,kp1:m) = CS(:,kp1:m)*HHk;
H(kp1:m,:) = HHk*H(kp1:m,:);
%The column has been reduced.
%The matrix is banded above k.
rowvec = SB(k1,kp1:m); Z = max(abs(rowvec));
                                   if k < m-2
nm = norm(SB(k1,kp2:m),inf);
    if nm < myzero
%The row does not have to be reduced.
        SB(k1,kp2:m) = zeros(1,m-kp1);
        k1 = k1+1;
    else
%Row k1 is reduced beyond kp2 with
 a HH transformation .
vk = SB(k1,kp2:m);
```

```
    if vk(1) == 0
        sgn = 1
    else
        sgn = sign(vk(1));
    end
u1 = sgn*norm(vk,2);
uk = [u1 zeros(1,m-kp2)]; wk = uk + vk; nmk = wk*wk';
HHk = eye(m-kp1) - 2*(wk'*wk)/nmk;
SB(k1:m,kp2:m) = SB(k1:m,kp2:m)*HHk;
CS(:,kp2:m) = CS(:,kp2:m)*HHk;
H(kp2:m,:) = HHk*H(kp2:m,:);
SB(kp2:m,kp1:m) = HHk*SB(kp2:m,kp1:m);
SB(k1,kp3:m) = zeros(1,m-kp3+1); %To eliminate
 roundoff error.
dm = SB(k1,kp1);
            if abs(dm) > sqrt(myzero)
tolk = abs(SB(k1,kp2)/dm);
             if tolk > tol
%Element SB(k1,kp2) cannot be reduced with
 a gaussian transformation.
             else
%We reduce SB(k1,kp2) to zero with a gaussian
 transformation.
w = SB(k1,kp2)/dm;


SB(k1+1:m,kp2) = SB(k1+1:m,kp2) - w*SB(k1+1:m,kp1);
SB(k1,kp2) = 0;
SB(kp1,kp1:m) = SB(kp1,kp1:m) + w*SB(kp2,kp1:m);
CS(:,kp2) = CS(:,kp2) - w*CS(:,kp1);
H(kp1,:) = H(kp1,:) + w*H(kp2,:);
k1 = k1+1;
             end
            end
    end
                        end %on k < m-2
                end %on k =1:m-2
        if k1 == m-2
dm = SB(k1,kp1);
            if abs(dm) > sqrt(myzero)
tolk = abs(SB(k1,kp2)/dm);
             if tolk > tol
%Element SB(k1,kp2) cannot be reduced with
 a gaussian transformation.
             else
```

```
%We reduce SB(k1,kp2) to zero with a gaussian
 transformation.
w = SB(k1,kp2)/dm;
SB(k1+1:m,kp2) = SB(k1+1:m,kp2) - w*SB(k1+1:m,kp1);
SB(k1,kp2) = 0;
SB(kp1,kp1:m) = SB(kp1,kp1:m) + w*SB(kp2,kp1:m);
CS(:,kp2) = CS(:,kp2) - w*CS(:,kp1);
H(kp1,:) = H(kp1,:) + w*H(kp2,:);
k1 = k1+1;
            end
           end
        end
k = m-1;
%SB is snorm*H*B*H^-1, and CS = snorm*G*C*H^-1
disp('1-norm of SB is')
norm(SB,1)
disp('number of elements beyond tridiagonal')
xtra = nnz(triu(SB,2))
errY = input('errY =')
    if isempty(errY)
        errY = 1e-6
    end
%We now compute the ADI iteration parameters.

parsyl
%We now iterate
adisyl
    if Yerr > 5*errY
%We cycle through the iterations one more time.
        adisyl
    end
    X = GI*Y*H;
    Z = A*X;
    W = X*B;
    Xerr = norm(C - Z - W,1)/norm(C,1);
        if Xerr > 10*errY
%We cycle through the iterations one more time.
            adisyl
    X = GI*Y*H;
    Z = A*X;
    W = X*B;
        end
disp('The estimated error is:')
    Xerr = norm(C - Z - W,1)/norm(C,1)
        if Xtr ~= zeros(nA,mB)
```

```
            disp('The true error is:')
            norm(X-Xtr,1)/norm(Xtr,1)
        end
return
************************************************
************************************************
************************************************
%adisyl performs the ADI iteration to solve
%the reduced Sylvester equation.
n = length(T); m = length(TB);
    if isempty(usol)
        usol = zeros(n,m);
    end
%usol is the initial estimate.
    if isempty(CS)
        error('no source term')
    end
    for j = 1:J
        rhsj = ws(j)*usol - usol*TB + CS;
        M = ws(j)*speye(n) + T;
        usol = M\rhsj;
        usp = usol';


        rhsj = wt(j)*usp - usp*T' + CS';
        M = wt(j)*speye(m) + TB';
        usol = (M\rhsj)';
    end
usol = real(usol);
Cest = T*usol + usol*TB;
disp('Estimated ||error in Y||/||Y||')
Yerr = norm(Cest - CS,1)/norm(CS,1)
Y = usol;
return
```

6.7 Solution of Low-Rank Lyapunov Equations with QGLOW

When C is symmetric and of low rank, QGLOW solves the Lyapunov equation with the Li–White algorithm. The reduction to banded form proceeds as described in Sect. 6.4. Iteration parameters are computed with QGPAR as for full-rank equations. The Li–White algorithm is performed in ADILOW.

```
%qglow solves for X the Lyapunov equation AX + XA'=C
 where eigenvalues
%of A (of order n) are in the right half plane
 bounded away from the imaginary axis.
%C = cl*cl' with cl of order nxm with m<n.
%A is reduced with a similarity transformation to a
 banded upper
%Hessenberg matrix S = snorm*G*A*inv(G). A matrix
 CS = snorm*G*cl is also generated.
%The value of snorm is the reciprocal of the
 square root
%of the product of the 1-norm and infinity-norm
 of matrix A. This bounds magnitudes
%of eigenvalues of S by unity. The reduced Lyapunov
 equation is SY + YS' = B where
%B = CS*CS'. Eigenvalues of S are determined
with eigs. ADI iteration parameters are
%computed with qgpar The Li-White ADI iteration
is performed with adilow
%to achieve a prescribed accuracy.  The inverse GI
of G is also generated for
%recovery of X = GI*Y*GI'/snorm.
%This m-file was generated on 12/31/12.
randn('seed',0)
A = input('Matrix A is ');
if isempty(A)
disp('A = randn(30) + s*eye(30) to assure
  N-stability')
A = randn(30);
A = A + (.3 + abs(min(real(eig(A)))))*eye(30);
%We require an N-stable matrix A.
end
n = length(A);
cl = input('Matrix cl is ');
if isempty(cl)
disp('cl = rand(n,2)')
cl = rand(n,2);
end
C = cl*cl';
if length(C) ~= n
disp('size of cl and A not consistent')
return
end
snorm = 1/sqrt(norm(A,1)*norm(A,inf));
k1 = 1;
```

```
S = snorm*A;
%This normalizes S so its eigenvalue magnitudes
are bounded by unity.
CS = sqrt(snorm)*cl;
GI = eye(n);
band =linspace(n,n,n);
tol = input('Bound on gaussian multipliers is:');
if isempty(tol)
    tol = 1e3
end
%The 1-norm of S is initially around 1.
Eigenvalues of the reduced
%matrix are reasonably
close to those of S even when tol is large.
                    for k = 1:n-2
kp1 = k+1; kp2 = k+2;kp3 = min(k+3,n);
colvec = S(kp1:n,k); Y = max(abs(colvec));
%The column is reduced with a HH transformation:
vk = S(kp1:n,k);
    if vk(1) == 0
        sgn = 1
    else
        sgn = sign(vk(1));
    end

u1 = sgn*norm(vk,2);
uk = [u1;zeros(n-kp1,1)]; wk = uk + vk; nmk = wk'*wk;
HHk = eye(n-k) - 2*(wk*wk')/nmk;
S(kp1:n,k:n) = HHk*S(kp1:n,k:n);
S(k1:n,kp1:n) = S(k1:n,kp1:n)*HHk;
CS(kp1:n,:) = HHk*CS(kp1:n,:);
S(kp2:n,k) = zeros(n-kp1,1); %To eliminate
roundoff error.
%The column has been reduced.  The matrix
is banded above k.
GI(:,kp1:n) = GI(:,kp1:n)*HHk; rowvec =
S(k1,kp1:n); Z = max(abs(rowvec));
                            if k < n-2
nm = norm(S(k1,kp2:n),inf);
    if nm < 1e-10
%The row does not have to be reduced.
        S(k1,kp2:n) = zeros(1,n-kp1);
        k1 = k1+1; band(k1) = kp1;
    else
%Row k1 is reduced beyond kp2 with
```

```
a HH transformation .
vk = S(k1,kp2:n);
    if vk(1) == 0
        sgn = 1
    else
        sgn = sign(vk(1));
    end
u1 = sgn*norm(vk,2);
uk = [u1 zeros(1,n-kp2)]; wk = uk + vk; nmk = wk*wk';
HHk = eye(n-kp1) - 2*(wk'*wk)/nmk;
S(k1:n,kp2:n) = S(k1:n,kp2:n)*HHk;
S(kp2:n,kp1:n) = HHk*S(kp2:n,kp1:n);
S(k1,kp3:n) = zeros(1,n-kp3+1);
To eliminate roundoff error.
GI(:,kp2:n) = GI(:,kp2:n)*HHk;
CS(kp2:n,:) = HHk*CS(kp2:n,:);
band(k1) = kp2;
    end
                                end %on k < n-2
band(k1) = kp2;
dm = S(k1,kp1);
           if abs(dm) > 1e-5
tolk = abs(S(k1,kp2)/dm);
            if tolk > tol

Element S(k1,kp2) cannot be reduced with
a gaussian transformation.
            else
%We reduce S(k1,kp2) to zero with
 a gaussian transformation.
w = S(k1,kp2)/dm;
S(k1+1:n,kp2) = S(k1+1:n,kp2) - w*S(k1+1:n,kp1);
S(k1,kp2) = 0;
GI(:,kp2) = GI(:,kp2) - w*GI(:,kp1);
S(kp1,kp1:n) = S(kp1,kp1:n) + w*S(kp2,kp1:n);
CS(kp1,:) = CS(kp1,:) + w*CS(kp2,:);
band(k1) = kp1; k1 = k1+1;
            end
           end
T = sparse(S);
                        end %on k =1:n-2
k = n-1;
sband = band;
disp('1-norm of the reduced matrix S is')
norm(T,1)
```

```
figure
spy(S);
hold on
title('Figure 1:  Profile of matrix S')
ylabel('matrix row'); xlabel('nonzero elements')
hold off
disp('number of elements beyond tridiagonal')
xtra = nnz(triu(S,2))
%ADI iteration parameters are determined.
errY = input('Bound on error in Y is ')
    if isempty(errY)
        errY = 1e-6;
    end
FCS = CS*CS';
qgpar
adilow
X = GI*Y*GI';
disp('Estimated ||error in X||/||X||')
errX = norm(C - A*X - X*A',1)/norm(C,1)
return
***********************************************
***********************************************
***********************************************
%adilow is preceded by qglow and qgpar.
%A low rank Lyapunov equation is solved with
Li-White ADI iteration
[n,m] = size(CS);
Z = zeros(n,m);
Y = zeros(n);
w = ww(1);
rtw = sqrt(w+conj(w));
RS = rtw*CS; %RS is the right-hand side for j = 1.
M = T + w*speye(n);
Z = M\RS;
Y = real(Z*Z');
%Y is the result of the first ADI iteration.
            for j = 2:J
wj = ww(j);
M = T + wj*speye(n);
rtwj = sqrt(wj+conj(wj));
rt = rtwj/rtw;
rtw = rtwj;
RS = rt*Z;    %This is the right-hand side
for iteration j.
M = T + wj*speye(n); wpwj = conj(wj) + wj; wp =
```

```
conj(w) + wj; w = wj; Z = M\RS; Z = RS - wp*Z;
%Z is the result of iteration j.
Y = Y + real(Z*Z');
            end
disp('Estimated ||error in Y||/||Y||')
errinY = norm(FCS - T*Y - Y*T',1)/norm(FCS,1)
return
```

6.8 Partitioning into a Sum of Low-Rank Matrices with LANCY

An SPD matrix may be partitioned to prescribed accuracy into a sum of low-rank matrices with the LANCY program. The Lanczos algorithm in LANCY allows choice of the rank of the components.

```
%LANCY is the Lanczos algorithm with matrix G
%approximated by low rank H = VT*T*VT'. 12/26/11.
G = input('matrix G is:');
    if isempty(G)
        G1 = 10*rand(30); G = G1 + G1';
    end
if norm(G - G',1) > 1e-10
    error('G not symmetric')
end
n = length(G);
    lowbnd = input('low bound on elements of T:')
        if isempty(lowbnd)
            lowbnd = .001;
        end
Gin = G;
mm = input('maximum rank of W');
    if isempty(mm)
        mm = ceil(sqrt(n));
    end
Vk = zeros(n,mm);
Tk = zeros(mm);
cold = 0;
korder = 1;
H = zeros(n);
delEG = 1.0;
errG = input('Desired accuracy is:')
    if isempty(errG)
        errG = .001
    end
```

```
k = 1;
% k is the number of low rank matrices.
krank = 0;
%krank is the rank of matrix k.
EG = G;
V1 = ones(n,1)/sqrt(n);
%V1 is the initial Lanczos vector.
                    while delEG > errG
V = zeros(n,mm);
V(:,1) = V1;
j = 1;
bett = 1;
bet(1) = 1;
G = EG;
    while bett > lowbnd & j <= mm
        wv = G*V(:,j);

            if j ~=1
                wv = wv - bet(j)*V(:,j-1);
            end
        alp(j) = wv'*V(:,j);
        wv = wv - alp(j)*V(:,j);
        bet(j+1) = norm(wv);
        V(:,j+1) = wv/bet(j+1);
        bett = bet(j+1)/bet(1);
        j = j+1;
    end
    r = j - 1;
    V1 = V(:,j);
    T = zeros(r);
    T(1,1) = alp(1);
        if r > 1
            T(1,2) = bet(2);
                for j = 2:r-1
        T(j,j) = alp(j);
        T(j-1,j) = bet(j);
        T(j,j-1) = bet(j);
                end
            T(r-1,r) = bet(r);
            T(r,r-1) = bet(r);
            T(r,r) = alp(r);
        end
VT = V(:,1:r);
H = H + VT*T*VT';
```

```
EG = Gin - H;
korder(k) = r;
Vk(:,cold+1:cold+r) = VT;
[rold cold] = size(Vk);
Tk(1:r,1:r,k) = T;
k = k+1;
delEG = norm(EG,1)/norm(Gin,1);
                                end
G = Gin;
disp('norm(G -H,1)/norm(G,1) =')
delEG
return
```

References

Abramowitz M, Stegun IA (1964) Handbook of mathematical functions, NBS AMS - 55; NIST Handbook of mathematical functions. Cambridge University Press, 2010

Achieser NI (1967) Theory of approximation. Ungar, New York

Bagby T (1969) On interpolation of rational functions. Duke J Math 36:95–104

Bartels RH, Stewart GW (1972) Solution of the matrix equation $AX + XB = C$. Commun. ACM 15:820–826

Benner P, Li R-C, Truhar N (2009) On the ADI method for Sylvester equations. J Comput Appl Math 233(4):1035–1045

Birkhoff G, Varga RS, Young DM (1962) Alternating direction implicit methods. Advances in computers, vol 3. Academic, New York, pp 189–273

Buzbee BL, Golub GH, Nielson CW (1970) On direct methods for solving Poisson's equation. SIAM J Numer Anal 7:627–656

Byers R (1983) Hamilton and symplectic algorithms for the algebraic Riccati equation. Ph.D. Thesis, Cornell University

Cauer W (1958) Synthesis of linear communication networks. McGraw-Hill, New York

Cesari L (1937) Sulla risoluzione dei sistemi di equazioni lineari par aprossimazioni successive. Atti Accad. Naz. Lincei. Rend. Cl. Sci. Fis. Mat. Nat. VI S–25:422–428

Clayton AJ (1963) Further results on polynomials having least maximum modulus over an ellipse in the complex plane. UKAEA Memorandum AAEEW, M 348

Concus P, Golub G (1973) Use of fast direct methods for the efficient numerical solution of nonseparable elliptic equations. SIAM J Numer Anal 20:1103–1120

Copson ET (1935) Theory of functions of a complex variable. Clarendon Press, Oxford

Dax A, Kaniel S (1981) The ELR method for computing the eigenvalues of a general matrix. SIAM J Numer Anal 18:597–605

Diliberto SP, Strauss EG (1951) On the approximation of a function of several variables by a sum of functions of fewer variables. Pacific J Math 1:195–210

Dodds HL, Sofu T, Wachspress EL (1990) A nine point ADI model problem for solution of the diffusion equation. Trans Am Nucl Soc 62:298

Douglas J Jr (1962) Alternating direction methods for three space variables. Numerische Mathematik 4:41–63

Douglas J Jr, Rachford HH Jr (1956) On the numerical solution of heat conduction problems in two or three space variables. Trans Am Math Soc 82:421–493

D'Yakonov EG (1961) An iteration method for solving systems of linear difference equations. Soviet Math (AMS translation) 2:647

Ellner NS, Wachspress EL (1986) Alternating direction implicit iteration for systems with complex spectra. SIAM J. Numer Anal 28(3):859–870 (1991); also New ADI model problem applications. In: Proceedings of fall joint computer conference, IEEE Computer Society Press,

© Springer Science+Business Media New York 2013

1986 and Master's thesis of N. Saltzman (Ellner), on ADI parameters for some complex spectra, University of Tennessee, Knoxville, TN, 1987
Engeli M, Ginsburg T, Rutishauser H, Stiefel E (1959) Refined iterative methods for computation of the solution and the eigenvalues of self-adjoint boundary value problems. Mitteilungen aus dem Institut für angewandte Mathematik, vol 8. Birkhäuser, Basel
Fischer B, Freund R (1991) Chebyshev polynomials are not always optimal. J Approx Theor 65(3):261–272
Gastinel N (1962) Sur le meilleur choix des paramètres de sur-relaxation. Chiffres 5:109–126
Geist GA (1989) Reduction of a general matrix to tridiagonal form. ORNL/TM-10991
Geist GA, Lu A, Wachspress EL (1989) Stabilized Gaussian reduction of an arbitrary matrix to tridiagonal form. ORNL/TM-11089
Golub GH, O'Leary DP (1987) Some history of the conjugate gradient and Lanczos algorithms. UMIACS-TR-87-20, CS-TR-1859, University of Maryland Institute for Advanced Computer Studies and Computer Science Department
Golub GH, Van Loan C (1983) Matrix computations. Johns Hopkins University Press, New York
Golub GH, Nash S, Van Loan C (1979) A Hessenberg-Shur method for solution of the problem $AX + XB = C$. IEEE Trans Automat Contr AC24:909–913
Gonchar AA (1969) Zolotarev problems connected with rational functions. Math USSR Sbornik 7:623–635
Guilinger WH (1965) Peaceman-Rachford method for small mesh increments. J Math Anal Appl 11:261–277
Gutknecht MH (1983) On complex rational approximation. In: Werner H et. al (eds) Part I: The characterization problem in computational aspects of complex analysis. Reidel, Dordrecht, pp 79–101
Hare DE, Tang WP (1989) Toward a stable tridiagonalization algorithm for general matrices. University of Waterloo technical report CS-89-03, 1989
Howell GW (1994) Toward an efficient and stable determination of spectra of a general matrix and a more efficient solution of the Lyapunov equation. First world congress of nonlinear analysis, 1994
Hestenes M, Stieffel E (1952) Methods of conjugate gradients for solving linear systems. J Res NBS 49(6):409–436
Howell G, Geist G (1995) Direct reduction to a similar near-tridiagonal form, section 6. In: Elmaghraby A, Ammar A (eds) Parallel and distributed computing systems. ISCA, Raleigh, pp 426–432
Howell G, Diaa N (2005) Algorithm 841: BHESS: Gaussian reduction to a similar banded Hessenberg form, ACM transactions on Mathematical Software vol. 31 (1)
Hubert L, Meulman J, Heiser W (2000) Two purposes for matrix factorization: A historical appraisal. SIAM Rev 42(1):68–82
Hurwitz H (1984) Infinitesimal scaling: A new procedure for modeling exterior field problems. IEEE Trans 20:1918–1923
Istace MP, Thiran JP (1993) On the third and fourth Zolotarev problems in the complex plane. Math Comput
Léja F (1957) Sur certaines suits liées aux ensem le plan et leur application à la representation conforme. Ann. Polon. Math 4:8–13
Lehoucq RB, Sorensen DC (1996) Deflation techniques for an implicitly restarted Arnoldi iteration. SIAM J Matrix Anal Appl 17:789–821
Li J-R, White J (2002) Low rank solution of Lyapunov equations. SIAM J Matrix Anal Appl 24(1):260–280
Lu A, Wachspress EL (1991) Solution of Lyapunov equations by ADI iteration. Comput Math Appl 21(9):43–58; Wachspress EL (1987) Iterative solution of the Lyapunov matrix equation. Appl Math Ltrs (1):87–90
Lynch RE, Rice JR (1968) Convergence rates of ADI methods with smooth initial error. Math Comp 22(102):331–335

Manteuffel TA (1977) The Tchebychev iteration for nonsymmetric linear systems. Numer Math 28:307–327
Nachtigal NM, Reddy SC, Trefethen LN (1990) How fast are nonsymmetric matrix iterations? MIT Numerical Analysis Report, pp 90–92, 1990
(2010) NIST Handbook of mathematical functions, chapter 19 by B.C. Carlson and chapter 22 by W.P. Reinhardt and P.L. Walke. Cambridge University Press, Cambridge
Opfer G, Schober G (1984) Richardson's iteration for nonsymmetric matrices. Linear Algebra Appl 58:343–361
Osborne MR, Watson GA (1978) Nonlinear approximation problems in vector norms. In: Watson GA (ed) Proceedings of Biennial conference on numerical analysis, Dundee 1977. Lecture notes in mathematics, vol 630. Springer, Berlin
Penzl T (1999) A cyclic low-rank Smith method for large sparse Lyapunov equations. SIAM J Sci Comp 21(4):1401–1418
Peaceman DW, Rachford HH Jr (1955) The numerical solution of parabolic and elliptic differential equation. J SIAM 3:28–41
Rivlin TJ (1980) Best uniform approximation by polynomials in the complex plane. In: Cheney EW (ed) Approximation theory III. Academic, New York
Ruttan A (1985) A characterization of best complex rational appproximants in a fundamental case. Const Approx 1:230–243
Saltzman N (1987) ADI parameters for some complex spectra. University of Tennessee Master's Thesis
Smirnov VI, Lebedev NA (1968) Functions of a complex variable. MIT Press, Cambridge
Smith RA (1968) Matrix equation $XA + XB = C$. SIAM J Appl Math 16(1):198–201
Starke G (1989) Rational minimization problems in connection with optimal ADI-parameters for complex domains. Ph.D. thesis, Institute für Praktische Mathematik, Universität Karlsruhe, FRG
Starke G (1991) Optimal alternating direction implicit parameters for nonsymmetric systems of linear equations. SIAM J Numer Anal 28:1431–1445
Stephenson K, Sundberg C (1985) Level curves of inner functions. Proc London Math Soc III 51:77–94
Tang WP (1988) A stabilized algorithm for tridiagonalization of an unsymmetric matrix. University of Waterloo Technical Report CS-88-14
Todd J (1984) Applications of transformation theory: A legacy from Zolotarev (1847–1878). Approximation theory and spline functions. D.Reidel, Dordrecht, pp 207–245
Varga RS (1962) Matrix iterative analysis. Prentice Hall, Englewood Cliffs
Wachspress EL (1957) CURE: A generalized two-space-dimension multigroup coding for the IBM-704. Knolls Atomic Power Report KAPL-1724, AEC R&D Physics and Mathematics TID-4500, 13th edn. Knolls Atomic Power Lab, Schenectady
Wachspress EL (1962) Optimum alternating-direction-implicit iteration parameters for a model problem. J Soc Indust Appl Math 10(2):339–350
Wachspress EL (1963) Extended application of alternating-direction-implicit iteration model problem theory. J Soc Indust Appl Math 11:994–1016
Wachspress EL (1966) Iterative solution of elliptic systems. Prentice Hall, Englewood Cliffs
Wachspress EL (1984) Generalized ADI preconditioning. Comput Math Appl 10(6):457–461
Wachspress EL (1988a) ADI solution of the Lyapunov equation. MSI Workshop on practical iterative methods for large-scale computations, Minneapolis, MN, 1988
Wachspress EL (1988b) Iterative solution of the Lyapunov matrix equation. Appl Math Lett 1: 87–90
Wachspress EL (1988c) The ADI minimax problem for complex spectra. Appl Math Lett 1(3):311–314
Wachspress EL (1990) The ADI minimax problem for complex spectra. In: Kincaid D, Hayes L, (eds) Iterative methods for large linear system. Academic, New York, pp 251–271

Wachspress EL (1991) Optimum alternating direction implicit iteration parameters for rectangular regions in the complex plane. Center for Mathematical Analysis of the Australian National University Report CMA-R33-90, 1991

Wachspress EL (1994) Three-variable alternating- direction- implicit iteration. Comput Math Appl 27:1–7

Wachspress EL (1995) The ADI model problem. Windsor, CA

Wachspress EL, Habetler GJ (1960) An alternating-direction-implicit iteration technique. J Soc Indust Appl Math 8(2):403–424

Wachspress, EL (2008) Trail to a Lyapunov equation solver, Computers and Mathematics with Applications 55, pp. 1653–1659

Watkins D (1988) Use of the LR algorithm to tridiagonalize a general matrix. SIAM annual meeting, Minneapolis, MN, 1988

Wilkinson JH (1965) The algebraic eigenvalue problem. Oxford University Press, Oxford

Wrigley HE (1963) Accelerating the Jacobi method for solving simultaneous equations by Chebyshev extrapolation when the eigenvalues of the iteration matrix are complex. Comput J 6:169–176

Young DM, Wheeler MF (1964) Alternating direction methods for solving partial difference equations. Nonlinear problems in engineering. Academic, New York, pp 220–246

Pemerburiskoi, Z, Oeuvres de G. Zolotarev (1877) Akademuu Nauk XXX, (5), pp 1–59

Index

E. Wachspress, *The ADI Model Problem*, DOI 10.1007/978-1-4614-5122-8,
© Springer Science+Business Media New York 2013

L

M

N

O

P

Printed in the United States
By Bookmasters

Printed in the United States
By Bookmasters